DISCRETE MATHEMATICS AND ITS APPLICATIONS

Series Editor KENNETH H. ROS

T0251596

PEARLS OF DISCRETE MATHEMATICS

Martin Erickson

Truman State University

Kirksville, Missouri, U.S.A.

CRC Press
Taylor & Francis Group
Boca Raton London New York

CRC Press is an imprint of the
Taylor & Francis Group, an **informa** business

A CHAPMAN & HALL BOOK

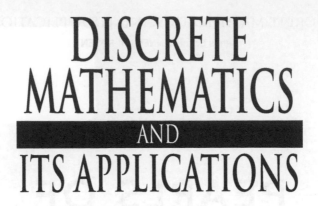

DISCRETE MATHEMATICS AND ITS APPLICATIONS

Series Editor

Kenneth H. Rosen, Ph.D.

Titles *(continued)*

CRC Press
Taylor & Francis Group
6000 Broken Sound Parkway NW, Suite 300
Boca Raton, FL 33487-2742

© 2010 by Taylor and Francis Group, LLC
CRC Press is an imprint of Taylor & Francis Group, an Informa business

No claim to original U.S. Government works

Printed in the United States of America on acid-free paper
10 9 8 7 6 5 4 3 2 1

International Standard Book Number: 978-1-4398-1616-5 (Paperback)

Library of Congress Cataloging-in-Publication Data

Erickson, Martin J., 1963-
    Pearls of discrete mathematics / Martin Erickson.
        p. cm. -- (Discrete mathematics and its applications)
    Includes bibliographical references and index.
    ISBN 978-1-4398-1616-5 (pbk. : alk. paper)
    1. Combinatorial analysis. 2. Graph theory. 3. Number theory. I. Title. II. Series.

QA164.E745 2009
511'.1--dc22   2009027565

Visit the Taylor & Francis Web site at
http://www.taylorandfrancis.com

and the CRC Press Web site at
http://www.crcpress.com

To my mother, Lorene Erickson

Contents

Preface

This book presents subjects suitable for a topics course or self-study in discrete mathematics. The focus is on representative, intriguing, and beautiful examples, problems, theorems, and proofs. Of course, the choice of coverage is personal and subjective, but I hope that the concepts I treat will be of interest to you and that you will find some as fascinating as I do.

Discrete structures (which are based on finite sets) comprise an area of mathematics that most people relate to naturally. How many ways can you choose an appetizer, a main course, and a dessert at dinner? How many walking routes can you take through town? How many lottery number combinations can you buy? These are typical counting problems in discrete mathematics. In our day-to-day mathematical lives, we often encounter counting problems or other types of problems involving discrete structures. We should learn the methods used to solve these problems.

Discrete structures play a central role in mathematics. They are intimately related to algebra, geometry, number theory, and combinatorics, and these relationships are illustrated with several of the pearls in this book. One needs only to look at the many journal titles in discrete mathematics (at least thirty in number) to see that this area is huge. The journal titles indicate connections between discrete mathematics and computing, information theory and codes, and probability. I think it's safe to say that all mathematicians and computer scientists would benefit from investigating the basic principles of discrete mathematics.

The world of discrete mathematics is like a mosaic or tapestry, with one pattern fitting into another and theories gradually emerging. I have taken an organic approach in this book, exploring concrete problems, introducing theory, and adding generalizations as we go.

Having taught mathematics for twenty-five years, I understand the importance of examples and exercises. Accordingly, I have kept examples and exercises at the forefront of the discussion. I've tried to arrange it so that each chapter features a particularly surprising, stunning, elegant, or unusual result. Included are mathematical items that don't appear in many books, such as the upward extension of Pascal's triangle, a recurrence relation for powers of Fibonacci numbers, the number of ways to make change for a million dollars, integer triangles, the period of Alcuin's sequence, Rook and Queen paths and the equivalent Nim and Wythoff's Nim games, the probability of a perfect bridge hand, random tournaments, a Fibonacci-like sequence of composite numbers, Shannon's theorems of information theory, higher-dimensional tic-tac-toe, animal achievement and avoidance games, and an algorithm for solving Sudoku puzzles and polycube packing problems. I introduce each chapter with a mathematical "teaser" or two to whet your appetite—mathematics can be engaging, inspiring, and even fun!

You will profit from doing the exercises, as a good deal of the mathematics is revealed there. The problems range in difficulty from easy to quite challenging. Exercises designated with a star ($\star$) are particularly difficult or require advanced mathematical background; exercises designated with a diamond ($\diamond$) require the use of a calculator or computer; exercises designated with a dagger ($\dagger$) are of theoretical importance. Hints or solutions to the exercises are provided in an appendix in the back of the book.

Thanks to the people who have kindly provided suggestions concerning this work: Robert Cacioppo, Robert Dobrow, Rodman Doll, Christine Erickson, Suren Fernando, David Garth, Joe Hemmeter, Daniel Jordan, Ken Price, Khang Tran, and Anthony Vazzana. Special thanks to Lorene Erickson for creating the cover artwork, *Spacescape V*. I would also like to thank the people at CRC Press, especially David Grubbs and Kenneth Rosen, for their help and encouragement in writing this book.

Part I

Counting: Basic

Chapter 1

Subsets of a Set

Abby has collected 100 *pennies. She offers Betty the choice of any or all of the pennies from her collection. The number of selections of pennies that Betty can make is* 1,267,650,600,228,229,401,496,703,205,376.

We begin with one of the simplest theorems of discrete mathematics. Denote by **N** the set of natural numbers, $\{1, 2, 3, \ldots\}$.

Theorem 1.1. Given $n \in \mathbf{N}$, a set with n elements has 2^n subsets.

Proof. Each element in the given n-element set can either be included or not included in a subset. Hence, there are

$$\underbrace{2 \times 2 \times 2 \times \cdots \times 2}_{n} = 2^n$$

choices in forming subsets. ∎

Notice that when choices are made independently, we *multiply* the numbers of choices. This principle is called the "product rule."

Example 1.2. How many subsets does the set $\{A, B, C\}$ have?

Solution: The set $\{A, B, C\}$ has $2^3 = 8$ subsets:

$$\emptyset, \{A\}, \{B\}, \{C\}, \{A, B\}, \{A, C\}, \{B, C\}, \{A, B, C\}.$$

∎

Example 1.3. How many subsets does a 100-element set have?

Solution: As stated in the teaser above, the number of subsets of a 100-element set is the colossal number

$$2^{100} = 1267650600228229401496703205376 \doteq 1.3 \times 10^{30}.$$

∎

Exercises

1. How many subsets does the set $\{A, B, C, D\}$ have? List the subsets.

2. At a certain dinner, there are four choices for the appetizer, two choices for the main course, and three choices for the dessert. How many different meals are possible?

♢3. Find the smallest integer n such that the number of subsets of an n-element set is greater than 10^{100} (a googol).

4. A teacher making a book display wants to showcase a novel, a history book, and a science book. There are four choices for the novel, two choices for the history book, and 10 choices for the science book. How many choices are possible for the three books?

5. A license consists of three digits (0 through 9), followed by a letter (A through Z), followed by another digit. How many different licenses are possible?

6. How many strings of 10 symbols are there in which the symbols may be 0, 1, or 2?

7. How many subsets of the set $\{a, b, c, d, e, f, g, h, i, j\}$ do not contain both a and b?

8. How many binary strings of length 99 are there such that the sum of the elements in the string is an odd number?

9. How many functions map the set $\{a, b, c\}$ to the set $\{w, x, y, z\}$?

10. How many functions map an n-element set to itself?

11. Let $X = \{1, 2, 3, \ldots, 2n\}$. How many functions map X to X such that each even number is mapped to an even number and each odd number is mapped to an odd number?

12. Is the result of Theorem 1.1 true for $n = 0$?

13. How many ways can you place a White King and a Black King on an 8×8 chessboard so that they don't attack each other? (A King attacks the squares horizontally, vertically, and diagonally adjacent to its own square.)

Chapter 2

Pascal's Triangle

Using Pascal's triangle, it is easy to calculate the number of subsets of a given set that have a certain size.

A *permutation* of a set is a selection of the elements of the set in some order. The number of permutations of n objects is $n!$, that is, the *factorial function* defined by

$$n! = n(n-1)(n-2)\ldots3\cdot2\cdot1, \text{ for } n \geq 1, \text{ and } 0! = 1.$$

Example 2.1. How many ways may 10 books be arranged on a shelf?

Solution: The number of arrangements is the number of permutations of 10 elements:

$$10! = 10\cdot9\cdot8\cdot7\cdot6\cdot5\cdot4\cdot3\cdot2\cdot1 = 3628800.$$

■

More generally, the number of permutations of k objects from a set of n objects is

$$P(n,k) = n(n-1)(n-2)\ldots(n-k+1) = \frac{n!}{(n-k)!}, \quad 0 \leq k \leq n.$$

Example 2.2. How many ways can four books from a set of 10 books be arranged on a shelf?

Solution: The number of arrangements is $P(10,4) = 10\cdot9\cdot8\cdot7 = 5040$. ■

A *combination* from a set is a subset of the set, i.e., a selection of elements of the set in which the order of the selected elements doesn't matter. The number of k-element combinations from an n-element set is

$$C(n,k) = \frac{P(n,k)}{k!} = \frac{n!}{k!(n-k)!}, \quad 0 \leq k \leq n.$$

We set $C(n,k) = 0$ if $k < 0$ or $k > n$.

The numbers $C(n,k)$ are given in Pascal's triangle, named after Blaise

$$1$$
$$1 \quad 1$$
$$1 \quad 2 \quad 1$$
$$1 \quad 3 \quad 3 \quad 1$$
$$1 \quad 4 \quad 6 \quad 4 \quad 1$$
$$1 \quad 5 \quad 10 \quad 10 \quad 5 \quad 1$$
$$1 \quad 6 \quad 15 \quad 20 \quad 15 \quad 6 \quad 1$$
$$1 \quad 7 \quad 21 \quad 35 \quad 35 \quad 21 \quad 7 \quad 1$$
$$1 \quad 8 \quad 28 \quad 56 \quad 70 \quad 56 \quad 28 \quad 8 \quad 1$$
$$1 \quad 9 \quad 36 \quad 84 \quad 126 \quad 126 \quad 84 \quad 36 \quad 9 \quad 1$$
$$1 \quad 10 \quad 45 \quad 120 \quad 210 \quad 252 \quad 210 \quad 120 \quad 45 \quad 10 \quad 1$$
$$\vdots$$

FIGURE 2.1: Pascal's triangle.

Pascal (1623–1662). See Figure 2.1. The rows of Pascal's triangle are numbered 0, 1, 2, etc. from top to bottom, and the columns are numbered 0, 1, 2, etc. from left to right. Each non-1 entry of Pascal's triangle is equal to the sum of the two entries directly above it.

The entry in row n, column k of Pascal's triangle is $C(n, k)$. This number is the same as the *binomial coefficient*

$$\binom{n}{k}.$$

For example, entry 2 of row 6 is $C(6,2) = \binom{6}{2} = 6!/(2!4!) = 15$.

Theorem 1.1 tells us that the sum of the numbers in row n of Pascal's triangle is 2^n. For example, the sum of the entries in the sixth row is $1 + 6 + 15 + 20 + 15 + 6 + 1 = 64 = 2^6$.

The simple rule that generates Pascal's triangle is a recurrence relation known as *Pascal's identity*:

$$\binom{n}{k} = \binom{n-1}{k-1} + \binom{n-1}{k}, \quad 1 \le k \le n-1, \quad \binom{n}{0} = 1, \; \binom{n}{n} = 1.$$

Pascal's identity has a counting proof. The binomial coefficient $\binom{n}{k}$ is the number of k-element subsets of the set $\{1, \ldots, n\}$. Each such subset either contains the element n or does not contain n. The number of k-element subsets that contain n is $\binom{n-1}{k-1}$. The number of k-element subsets that don't contain n is $\binom{n-1}{k}$.

Binomial coefficients get their name because they are the coefficients of a binomial expansion. That is, $\binom{n}{k}$ is the coefficient of $a^{n-k}b^k$ in $(a+b)^n$. The binomial theorem says just that.

Theorem 2.3 (Binomial Theorem).

$$(a+b)^n = \sum_{k=0}^{n} \binom{n}{k} a^{n-k}b^k, \quad n \geq 0.$$

Example 2.4. Give the expansion of $(a+b)^6$.

Solution: By the binomial theorem, the expansion of $(a+b)^6$ is

$$\binom{6}{0}a^6b^0 + \binom{6}{1}a^5b^1 + \binom{6}{2}a^4b^2 + \binom{6}{3}a^3b^3 + \binom{6}{4}a^2b^4 + \binom{6}{5}a^1b^5 + \binom{6}{6}a^0b^6$$

$$= a^6 + 6a^5b + 15a^4b^2 + 20a^3b^3 + 15a^2b^4 + 6ab^5 + b^6.$$

∎

Armed with the binomial theorem, we can give a second proof of Theorem 1.1. Set $a = 1$ and $b = 1$, and we get

$$2^n = \sum_{k=0}^{n} \binom{n}{k}.$$

Thus, there are 2^n subsets of an n-element set.

Exercises

1. A teacher has eight books to put on a shelf. How many different orderings of the books are possible?

2. A student has 10 books but only room for six of them on a shelf. How many permutations of the books are possible on the shelf?

◇3. Find the smallest integer n such that $n! > 10^{100}$ (a googol).

4. A couple plans to visit three selected cities in Germany, followed by four selected cities in France, followed by five selected cities in Spain. In how many ways can the couple order their itinerary?

5. You have three small glasses, four medium-size glasses, and five large glasses. If glasses of the same size are indistinguishable, how many ways can you arrange the glasses in a row?

6. A librarian wants to arrange four astronomy books, five medical books, and six religious books on a shelf. Books of the same category should be grouped together, but otherwise the books may be put in any order. How many orderings are possible?

7. In how many ways can you arrange the letters of the word RHODO-DENDRUM?

8. How many one-to-one functions are there from the set $\{a, b, c\}$ to the set $\{t, u, v, w, x, y, z\}$?

9. Let X be an n-element set, where $n \in \mathbf{N}$. How many functions from X to X are not one-to-one?

10. Find a formula for the number of different binary relations possible on a set of n elements, where $n \in \mathbf{N}$.

11. Professor Bumble teaches five different classes, A, B, C, D, and E. He prepared a different lecture for each class today, but he gave some or all of the lectures to the wrong classes. He knows that class A received the wrong lecture. In how many different orders can Professor Bumble have given his lectures today?

12. A singer plans to perform three songs from a repertoire of 12 songs. How many different programs are possible if there are two songs, A and B, that cannot both be performed?

13. A student decides to take three classes from a set of 10. In how many ways may she do this?

14. In a certain lottery, a contestant must choose six numbers from the set 1, 2, ..., 44. How many combinations are possible?

15. Evaluate $\binom{20}{10}$.

16. What is the coefficient of $a^{10}b^{10}$ in the expansion of $(a + b)^{20}$?

17. Give the expansion of $(a + b)^{10}$.

18. Give simple formulas for $\binom{n}{2}$ and $\binom{n}{3}$.

19. What is the coefficient of x^{10} in $(1 + x^2)^{20}$?

20. What is the coefficient of x^7 in $(1 - 2x)^{10}$?

21. What is the constant in the expansion of $(x + 1/x)^{20}$?

22. What is the constant in the expansion of $(x^4 + 1/x)^{20}$?

23. Explain the formula $C(n, k) = P(n, k)/k!$.

24. Give an algebraic proof of Pascal's identity.

†25. Prove the binomial theorem.

26. Professor Bumble doesn't remember how many students are in his honors mathematics class. But he does remember that there are 924 ways to divide the class into two equal-size groups of students. Help Professor Bumble determine how many students are in his class.

27. A pointer starts at 0 on the real number line and moves right or left one unit at each step. Let n and k be positive integers. How many different paths of k steps terminate at the integer n?

◇28. Use the recurrence relation for Pascal's triangle to compute the value of the binomial coefficient $\binom{100}{50}$.

29. Suppose that n lines are given in the plane in "general position" (no two parallel and no three concurrent at a point). Into how many regions is the plane partitioned?

30. Suppose that n planes are given in three-dimensional space in "general position" (no two parallel, no three concurrent in a line, and no four concurrent in a point). Into how many regions is space partitioned?

Chapter 3

Binomial Coefficient Identities

Every finite nonempty set has as many subsets with an even number of elements as subsets with an odd number of elements.

Consider again the subsets of the three-element set $\{A, B, C\}$. Half of these subsets have an even number of elements and half an odd number of elements:

even number of elements	odd number of elements
$\emptyset$	$\{A\}$
$\{A, B\}$	$\{B\}$
$\{A, C\}$	$\{C\}$
$\{B, C\}$	$\{A, B, C\}$.

We claim that this property holds for every finite nonempty set.

Proposition 3.1. Given $n \geq 1$, half of the subsets of an n-element set have an even number of elements and half have an odd number of elements.

Proof. Let $a = 1$ and $b = -1$ in the binomial theorem. Then

$$\sum_{k=0}^{n} \binom{n}{k}(-1)^k = (-1 + 1)^n = 0^n = 0, \quad n \geq 1.$$

Putting the summands associated with even k on one side of the relation and those associated with odd k on the other side, we obtain

$$\sum_{k \text{ odd}} \binom{n}{k} = \sum_{k \text{ even}} \binom{n}{k}.$$

■

For n odd, this assertion follows trivially from the symmetry of the binomial coefficients, that is, $\binom{n}{k} = \binom{n}{n-k}$. We give a counting argument valid for all $n \geq 1$. Let $X = \{1, 2, 3, \ldots, n\}$ and

$$
\begin{aligned}
\mathcal{A} &= \{S \subseteq X : |S| \text{ is even and } 1 \in S\} \\
\mathcal{B} &= \{S \subseteq X : |S| \text{ is odd and } 1 \in S\} \\
\mathcal{C} &= \{S \subseteq X : |S| \text{ is even and } 1 \notin S\} \\
\mathcal{D} &= \{S \subseteq X : |S| \text{ is odd and } 1 \notin S\}.
\end{aligned}
$$

There is a one-to-one correspondence between $\mathcal{A}$ and $\mathcal{D}$: simply remove 1 from every element of $\mathcal{A}$ to form an element of $\mathcal{D}$. (What is the inverse?) Similarly, there is a one-to-one correspondence between $\mathcal{B}$ and $\mathcal{C}$. Hence $|\mathcal{A}| = |\mathcal{D}|$ and $|\mathcal{B}| = |\mathcal{C}|$, and it follows that

$$|\mathcal{A}| + |\mathcal{C}| = |\mathcal{B}| + |\mathcal{D}|.$$

Let's explore a few more binomial coefficient identities.

Example 3.2. What is the sum $\sum_{k=0}^{n} \binom{n}{k}^2$?

Solution: We work out some examples using Pascal's triangle:

$$
\begin{aligned}
n = 1 &: \ 1^2 = 1 \\
n = 2 &: \ 1^2 + 1^2 = 2 \\
n = 3 &: \ 1^2 + 2^2 + 1^2 = 6 \\
n = 4 &: \ 1^2 + 3^2 + 3^2 + 1^2 = 20 \\
n = 5 &: \ 1^2 + 4^2 + 6^2 + 4^2 + 1^2 = 70.
\end{aligned}
$$

We recognize these sums as central binomial coefficients, and we make the conjecture that

$$\sum_{k=0}^{n} \binom{n}{k}^2 = \binom{2n}{n}.$$

Typically, the mathematical process consists of working examples, looking for patterns, making conjectures, and proving the conjectures. Let's try to prove our conjecture.

We rewrite our conjecture as follows:

$$\binom{n}{0}\binom{n}{0} + \binom{n}{1}\binom{n}{1} + \binom{n}{2}\binom{n}{2} + \cdots + \binom{n}{n}\binom{n}{n} = \binom{2n}{n}.$$

We know that the right side counts the ways of selecting n numbers from the set $\{1, 2, 3, \ldots, 2n\}$. Why is this counted by the left side? Rewrite a little, using symmetry:

$$\binom{n}{0}\binom{n}{n} + \binom{n}{1}\binom{n}{n-1} + \binom{n}{2}\binom{n}{n-2} + \cdots + \binom{n}{n}\binom{n}{0} = \binom{2n}{n}.$$

Now the truth of the identity is clear. The right side counts the number of n-element subsets of $\{1, 2, 3, \ldots, 2n\}$. The left side counts the same thing, where, for $0 \leq k \leq n$, the term $\binom{n}{k}\binom{n}{n-k}$ counts the number of subsets in which k elements are chosen from the set $\{1, \ldots, n\}$ and $n - k$ elements are chosen from the set $\{n + 1, \ldots, 2n\}$. ∎

The identity of Example 3.2 has a counting interpretation in terms of *lattice paths*, i.e., paths along the lines of an $n \times n$ grid. The binomial coefficient $\binom{2n}{n}$ is the number of northeast paths that start at the southwest corner of the grid and stop at the northeast corner. Each path has length $2n$ and is determined by a sequence of n "easts" and n "norths" in some order. The summation $\sum_{k=0}^{n} \binom{n}{k}^2$ counts the paths according to their intersection with the main southeast diagonal of the grid. The number of paths that cross the diagonal at the point k units east of the starting point is $\binom{n}{k}^2$, where $0 \leq k \leq n$.

Both Pascal's identity and the identity of Example 3.2 are generalized by *Vandermonde's identity*, credited to Alexandre-Théophile Vandermonde (1735–1796):

$$\binom{m+n}{k} = \sum_{i=0}^{k} \binom{m}{i}\binom{n}{k-i}, \quad m, n, k \geq 0.$$

The binomial coefficient $\binom{m+n}{k}$ is the number of k-element subsets of the set $\{1, \ldots, m+n\}$. The number of such subsets that contain i elements from the set $\{1, \ldots, m\}$ and $k-i$ elements from the set $\{m+1, \ldots, m+n\}$ is $\binom{m}{i}\binom{n}{k-i}$. The summation $\sum_{i=0}^{k} \binom{m}{i}\binom{n}{k-i}$ counts these subsets for $0 \leq i \leq k$. Letting $m = 1$, and changing n to $n - 1$, the relation becomes Pascal's identity. Letting $k = m = n$, we obtain the identity from Example 3.2.

The next identity is often useful (e.g., see Chapter 13).

Proposition 3.3 ("Subcommittee Identity"). For $0 \leq j \leq k \leq n$, we have

$$\binom{n}{k}\binom{k}{j} = \binom{n}{j}\binom{n-j}{k-j}.$$

Proof. Both expressions count the number of ways to choose, from n people, a committee of size k and a subcommittee of size j. ∎

Example 3.4. Prove the identity

$$\sum_{k=0}^{n} \binom{n+k}{n} 2^{-k} = 2^n.$$

Solution: We give a counting proof of the equivalent identity

$$\sum_{k=0}^{n} \binom{n+k}{n} 2 \cdot 2^{n-k} = 2^{2n+1}.$$

The right side of this relation is the number of binary strings of length $2n + 1$. We must show that the left side counts the same strings. Each binary string of length $2n + 1$ contains at least $n + 1$ 0s or at least $n + 1$ 1s (but not both). Counting from the left, let $n + k + 1$, where $0 \le k \le n$, be the position of the $(n + 1)$st 0 or $(n + 1)$st 1. There are two possibilities for this element (0 or 1); there are $\binom{n+k}{n}$ binary strings of length $n + k$ that contain n of one symbol and k of the other; and there are 2^{n-k} choices for the remaining $n - k$ elements. This establishes the identity. ∎

The binomial theorem generalizes to arbitrary exponents. For any real number α, and k a positive integer, define

$$\binom{\alpha}{k} = \frac{\alpha(\alpha - 1)(\alpha - 2)\ldots(\alpha - k + 1)}{k!}.$$

Also, define $\binom{\alpha}{0} = 1$. For a proof of the following theorem, see [Rud76].

Theorem 3.5 (Binomial Series). Let α be a real number and $|x| < 1$. Then

$$(1 + x)^\alpha = \sum_{k=0}^{\infty} \binom{\alpha}{k} x^k.$$

Example 3.6. Let n be an integer greater than or equal to 1. Prove the formula

$$\frac{1}{(1 + x)^n} = \sum_{k=0}^{\infty} (-1)^k \binom{n + k - 1}{k} x^k, \quad |x| < 1.$$

Solution: By the binomial series theorem,

$$(1 + x)^{-n} = \sum_{k=0}^{\infty} \binom{-n}{k} x^k.$$

The result now follows from the identity (see Exercises)

$$\binom{-n}{k} = (-1)^k \binom{n + k - 1}{k}.$$

∎

Example 3.7. Prove the identity

$$\sum_{k=1}^{n} k \binom{n}{k} = n2^{n-1}.$$

Solution: We give three proofs.

(1) The first proof is algebraic. Noting that we can "pull an n out" of each term in the sum, we obtain

$$\sum_{k=1}^{n} k \binom{n}{k} = \sum_{k=1}^{n} k \frac{n!}{k!(n-k)!}$$

$$= n \sum_{k=1}^{n} \frac{(n-1)!}{(k-1)!(n-k)!}$$

$$= n \sum_{k=1}^{n} \binom{n-1}{k-1}$$

$$= n2^{n-1}.$$

(2) The second proof is a counting proof. Consider all possible ways of choosing a team and a team leader from a group of n people. The left side clearly counts this, according to the size k of the team. The right side counts the same thing, as we have n choices for the leader and each other person can be on or off the team.

(3) The third proof uses calculus. From the binomial theorem, we have

$$(x+1)^n = \sum_{k=0}^{n} \binom{n}{k} x^k.$$

Taking a derivative with respect to x yields

$$n(x+1)^{n-1} = \sum_{k=1}^{n} \binom{n}{k} k x^{k-1}.$$

Evaluating both sides of the last relation at $x = 1$, we obtain our desired identity.

See Chapter 10 for a probabilistic proof of this result. ∎

Exercises

1. Prove the following identities:

   (a) $\binom{n}{k} = \frac{n}{k}\binom{n-1}{k-1}$

   (b) $\binom{-n}{k} = (-1)^k \binom{n+k-1}{k}$.

2. Prove the identity

$$1 \cdot 2 + 2 \cdot 3 + 3 \cdot 4 + \cdots + n \cdot (n+1) = n(n+1)(n+2)/3.$$

3. Find and prove a simple formula for

$$1 \cdot 2 \cdot 3 + 2 \cdot 3 \cdot 4 + 3 \cdot 4 \cdot 5 + \cdots + n \cdot (n+1) \cdot (n+2).$$

4. Prove the following identities:

   (a) $\binom{2n-1}{n} = \sum_{k=0}^{n} \binom{n}{k}\binom{n-1}{k}$

   (b) $\binom{3n}{n} = \sum_{k=0}^{n} \binom{n}{k}\binom{2n}{k}$.

5. (a) Prove the identity $\binom{n}{k} = \frac{n-k+1}{k}\binom{n}{k-1}$.

   (b) Use the identity of part (a) to show that the entries of each row of Pascal's triangle increase from left to right, attain a maximum value at the middle entry (or two middle entries), and then decrease.

6. Prove the inequality $\binom{n}{k}^2 \geq \binom{n}{k-1}\binom{n}{k+1}$, where $1 \leq k \leq n-1$.

7. Prove the identity

$$\binom{m-1}{k} = \sum_{i=0}^{k} \binom{m}{i}(-1)^{k-i}.$$

8. Suppose that five particles travel back and forth on the unit interval $[0, 1]$. At the start, all five particles move to the right with the same speed. When a particle reaches 0 or 1, it reverses direction but maintains its speed. When two particles collide, they both reverse direction (and maintain their speeds). How many particle–particle collisions occur before the particles once again occupy their original positions and are moving to the right?

9. Show that the number of ways that $2n$ people may be paired into n pairs is $\binom{2n}{n}n!2^{-n}$.

10. Prove that the number
$$\frac{4700!}{100!(47!)^{100}}$$

is an integer. Do this without actually calculating the number.

11. Prove the identity
$$\binom{m+n+1}{n+1} = \sum_{i=0}^{m} \binom{n+i}{n}.$$

12. Simplify the expression
$$\sum_{k=1}^{n} \binom{n}{k} \frac{k^2}{3^k}.$$

13. Prove that the number of binary strings of length n that contain exactly k copies of the string 10 is
$$\binom{n+1}{2k+1}.$$

†⋆14. For each integer $k \geq 0$, define
$$S_k(n) = \sum_{i=1}^{n} i^k.$$

Give formulas for $S_0(n)$, $S_1(n)$, $S_2(n)$, and $S_3(n)$. Prove that $S_k(n)$ is a polynomial in n of degree $k+1$ with leading coefficient $1/(k+1)$.

◇15. Use a computer to find $S_{10}(n)$, where this polynomial is defined in the previous exercise.

16. Give the first several terms of the expansion of $(1+x)^{-4}$ in powers of x.

◇17. Use a computer to give the first 10 terms of the expansion of $(1+3x)^{-7}$ in powers of x.

†18. Prove the multinomial theorem: In the expansion of
$$(x_1 + x_2 + \cdots + x_k)^n,$$

the coefficient of $x_1^{\alpha_1} x_2^{\alpha_2} \ldots x_k^{\alpha_k}$, where the α_i are nonnegative integers such that $\alpha_1 + \alpha_2 + \cdots + \alpha_k = n$, is the multinomial coefficient
$$\binom{n}{\alpha_1, \alpha_2, \ldots, \alpha_k} = \frac{n!}{\alpha_1! \alpha_2! \ldots \alpha_k!}.$$

19. Give the expansion of $(a + b + c)^4$.

20. What is the coefficient of $x^3 y^7$ in the expansion of $(x + y + 1)^{20}$?

21. Let $\mathbf{R}$ be the set of real numbers.

    (a) How many paths in $\mathbf{R}^2$ start at the origin $(0, 0)$, move in steps of $(1, 0)$ or $(0, 1)$, and end at $(10, 15)$?

    (b) How many paths in $\mathbf{R}^3$ start at the origin $(0, 0, 0)$, move in steps of $(1, 0, 0)$, $(0, 1, 0)$, or $(0, 0, 1)$, and end at $(10, 15, 20)$?

◇22. Use a computer and the multinomial theorem to find the coefficient of $x^{10} y^{10}$ in the expansion of $(1 + x + y)^{100}$.

23. You can order a pizza with up to four toppings (repetitions allowed) from a set of 12 toppings. The order of the toppings is unimportant. How many different pizzas can you order?

24. How many solutions does the equation

$$x_1 + x_2 + x_3 = 10$$

have in nonnegative integers?

†25. How many solutions does the equation

$$x_1 + \cdots + x_k = n$$

have in nonnegative integers?

26. How many ways can k indistinguishable balls be placed in n distinguishable urns so that each urn contains an odd number of balls?

27. Prove the identity

$$\sum_{j=0}^{n} \sum_{k=0}^{n} \binom{n + j + k}{n, j, k} 3^{-j-k} = 3^n.$$

†28. Let $S(n)$ be the number of ways that n can be written as a sum of positive integers: $n = n_1 + \cdots + n_k$ for any k (order important). Such summations are called *compositions* of n.

†29. Show that the number of permutations of n elements with an odd number of cycles is equal to the number of permutations of n elements with an even number of cycles.

Part II

Counting: Intermediate

Chapter 4

Finding a Polynomial

There is a simple method for finding a polynomial given its initial values.

Suppose that the values of a polynomial $p(n)$, for $n \geq 0$, are

$$5, \ 13, \ 25, \ 83, \ 277, \ 745, \ 1673, \ 3295, \ 5893, \ 9797, \ 15385, \ \dots.$$

What is $p(n)$?

We take successive differences of consecutive terms of the sequence, forming the first difference sequence:

$$8, \ 12, \ 58, \ 194, \ 468, \ 928, \ 1622, \ 2598, \ 3904, \ 5588, \ \dots.$$

Repeating this operation, we find the second difference sequence:

$$4, \ 46, \ 136, \ 274, \ 460, \ 694, \ 976, \ 1306, \ 1684, \ \dots.$$

The third difference sequence is

$$42, \ 90, \ 138, \ 186, \ 234, \ 282, \ 330, \ 378, \ \dots.$$

The fourth difference sequence is

$$48, \ 48, \ 48, \ 48, \ 48, \ 48, \ 48, \ \dots.$$

Having obtained a constant sequence, we stop. We find $p(n)$ by multiplying the first terms of these difference sequences by successive binomial coefficients:

$$p(n) = 5\binom{n}{0} + 8\binom{n}{1} + 4\binom{n}{2} + 42\binom{n}{3} + 48\binom{n}{4}$$

$$= 2n^4 - 5n^3 + 3n^2 + 8n + 5.$$

If $p(n)$ is a polynomial of degree 4, then we have found it. However, it's possible that $p(n)$ is only masquerading as this simple polynomial and more of its values would reveal a different (higher degree) answer.

Why does this method work? The key is Pascal's identity:

$$\binom{n}{k} = \binom{n-1}{k} + \binom{n-1}{k-1}, \quad 1 \leq k \leq n-1.$$

Every polynomial of degree d can be written as a linear combination of the polynomials $\binom{n}{k}$, for $0 \leq k \leq d$. Let's see what happens under the difference operations to the sequence given by the polynomial $\binom{n}{k}$. From Pascal's identity, the sequence

$$\binom{0}{k}, \binom{1}{k}, \binom{2}{k}, \binom{3}{k}, \ldots$$

yields the first difference sequence

$$\binom{0}{k-1}, \binom{1}{k-1}, \binom{2}{k-1}, \binom{3}{k-1}, \ldots.$$

Continuing in this manner, the kth difference sequence is

$$\binom{0}{0}, \binom{1}{0}, \binom{2}{0}, \binom{3}{0}, \ldots.$$

Since these terms are all equal to 1, we stop, and the contribution to $p(n)$ from our calculation is $1 \cdot \binom{n}{k} = \binom{n}{k}$, which is correct. Since $p(n)$ can be written as a linear combination of the polynomials $\binom{n}{k}$, and each such polynomial gives the correct contribution in our formula, then our formula is correct.

Example 4.1. Let's try another sequence, say,

$$0, 5, 18, 45, 92, 165, 270, 413, \ldots.$$

What polynomial produces this sequence?

Solution: We write down the array of difference sequences:

$$
\begin{array}{cccccccc}
0, & 5, & 18, & 45, & 92, & 165, & 270, & 413, \quad \ldots \\
5, & 13, & 27, & 47, & 73, & 105, & 143, & \quad \ldots \\
8, & 14, & 20, & 26, & 32, & 38, & \quad \ldots \\
6, & 6, & 6, & 6, & 6, & \quad \ldots .
\end{array}
$$

We obtain the polynomial

$$0\binom{n}{0} + 5\binom{n}{1} + 8\binom{n}{2} + 6\binom{n}{3} = n^3 + n^2 + 3n.$$

Exercises

1. Suppose that the sequence

    $$7,\ 11,\ 25,\ 73,\ 203,\ 487,\ 1021,\ 1925,\ 3343,\ 5443,\ 8417,\ \ldots$$

    represents the values of a polynomial $p(n)$, where $n = 0, 1, 2, \ldots$. What is the polynomial?

2. Professor Bumble wrote down the first few terms $p(0)$, $p(1)$, etc. of a polynomial,

    $$0,\ 1,\ 32,\ 243,\ 1024,\ 3125,\ 7776,\ 16807,\ 32768,\ 59049,\ 100000,\ \ldots,$$

    but he forgot what polynomial he started with. Help Professor Bumble find $p(n)$.

3. Show how to write the polynomial $n^3 + 2n^2 - n + 1$ as a linear combination of the polynomials $\binom{n}{k}$, for $0 \le k \le 3$.

4. Suppose that p is a polynomial such that $p(0) = 3$, $p(2) = 5$, $p(4) = 39$, $p(6) = 153$, and $p(8) = 395$. What is your best guess for $p(n)$?

†5. Explain why every polynomial of degree d can be written as a linear combination of the polynomials $\binom{n}{k}$, for $0 \le k \le d$.

†6. Prove that

    $$n^d = \sum_{k=1}^{d} T(d, k) \binom{n}{k},$$

    where $T(d, k)$ is the number of onto functions from $\{1, \ldots, d\}$ to $\{1, \ldots, k\}$.

◇7. Professor Bumble wrote the values of a polynomial in two variables, $p(m, n)$, for $0 \le m, n \le 5$, in a two-dimensional array, as follows.

10	16	22	28	34	40
15	23	33	45	59	75
20	36	68	116	180	260
25	61	151	295	493	745
30	104	306	636	1094	1680
35	171	557	1193	2079	3215

    Professor Bumble has forgotten what polynomial he started with. Can you devise a simple, fast way to find $p(m, n)$?

8. Find a recurrence relation for the sequence

$$3, 2, -1, 0, 11, 38, 87, 164, 275, 426, 623, \ldots.$$

9. Suppose that p is a polynomial of degree 3 such that $p(0) = 1$, $p'(0) = 1$, $p''(0) = 4$, and $p'''(0) = 18$. What is $p(n)$?

10. Is there a polynomial $p(n)$, with integer coefficients, whose values, for $n \geq 0$, are

$$1, 1, 2, 2, 3, 3, 4, 4, 5, 5, 6, 6, \ldots?$$

Chapter 5

The Upward-Extended Pascal's Triangle

Do you know that Pascal's triangle extends upward? The extended triangle gives the coefficients of binomial series for negative exponents.

In Figure 5.1, Pascal's identity is used to calculate binomial coefficients $\binom{n}{k}$ with negative values of n. (In the figure, the triangle is left-justified and some entries are padded with 0s to aid in the calculation.) The recurrence relation is

$$\binom{n}{k} = \binom{n+1}{k} - \binom{n}{k-1}, \quad k \geq 0,$$

and we define $\binom{n}{-1} = 0$, for all n. Try to verify some of the entries in the extended Pascal's triangle. Do you recognize the values? They are the numbers $(-1)^k \binom{n+k-1}{k}$, given by the binomial series theorem for the coefficients of x in the expansion of $(1+x)^{-n}$.

Example 5.1. Give the first several terms of the expansion of $(1+x)^{-4}$ in powers of x.

Solution: We locate the coefficients of the expansion in row -4 of the extended Pascal's triangle. Thus

$$(1+x)^{-4} = 1 - 4x + 10x^2 - 20x^3 + 35x^4 - 56x^5 + \cdots.$$

■

Exercises

1. Use the extended Pascal's triangle to give the first several terms of the expansion of $(1+x)^{-5}$ in powers of x.

2. Use the extended Pascal's triangle to give the first several terms of the expansion of $(1-x^2)^{-4}$ in powers of x.

⋮	⋮	⋮	⋮	⋮	⋮	⋮	
0	1	−5	15	−35	70	−126	⋯
0	1	−4	10	−20	35	−56	⋯
0	1	−3	6	−10	15	−21	⋯
0	1	−2	3	−4	5	−6	⋯
0	1	−1	1	−1	1	−1	⋯
0	1	0	0	0	0	0	⋯
0	1	1	0	0	0	0	⋯
0	1	2	1	0	0	0	⋯
0	1	3	3	1	0	0	⋯
0	1	4	6	4	1	0	⋯
0	1	5	10	10	5	1	⋯
⋮	⋮	⋮	⋮	⋮	⋮	⋮	

FIGURE 5.1: The extended Pascal's triangle.

3. Calculate by hand (using the recurrence relation) the first few terms of row −6 of Pascal's triangle.

4. What is the sum of the series

$$1 - \frac{2}{7} + \frac{3}{7^2} - \frac{4}{7^3} + \frac{5}{7^4} - \frac{6}{7^5} + \cdots ?$$

5. What is the sum of the series

$$1 + 3\left(\frac{2}{3}\right) + 6\left(\frac{2}{3}\right)^2 + 10\left(\frac{2}{3}\right)^3 + 15\left(\frac{2}{3}\right)^4 + 21\left(\frac{2}{3}\right)^5 + \cdots ?$$

6. If the first few terms of the expansion of $(1 + x)^n + (1 + x)^{-n}$, in powers of x, are

$$2 + 9x^2 - 9x^3 + 15x^4 - 21x^5 + 28x^6 - 36x^7 + 45x^8 - 55x^9 + \cdots ,$$

what is n?

7. Professor Bumble computes the first few terms of the expansion of $(1 + x)^n$, for some integer n. One of the terms he obtains is $15x^4$. Later, he forgets the value of n that he used. Help Bumble find the possible values of n.

◇8. Use a computer and Pascal's recurrence relation to generate a table of binomial coefficients $\binom{n}{k}$, with $-10 \le n \le 10$ and $0 \le k \le 10$.

Chapter 6

Recurrence Relations and Fibonacci Numbers

Infinitely many numbers appear at least six times in Pascal's triangle.

The kth powers of the Fibonacci numbers satisfy a linear homogeneous recurrence relation of order $k + 1$ with integer coefficients.

Let's consider one of the most famous sequences of numbers, the Fibonacci sequence, named after Leonardo of Pisa, a.k.a. Leonardo Fibonacci (1170–1250). The Fibonacci sequence $\{F_0, F_1, F_2, \ldots\}$ is defined recursively by the initial values

$$F_0 = 0, \; F_1 = 1,$$

and the recurrence relation

$$F_n = F_{n-1} + F_{n-2}, \quad \text{for } n \geq 2.$$

Thus, the Fibonacci numbers are

$$0, \; 1, \; 1, \; 2, \; 3, \; 5, \; 8, \; 13, \; 21, \; 34, \; 55, \; 89, \; 144, \; 233, \; 377, \; 610, \; \ldots.$$

Fibonacci numbers count many things. For example:

- F_{n+1} is the number of ways that an $n \times 1$ box may be packed with 2×1 and 1×1 boxes.

- F_{n+2} is the number of binary strings of length n that do not contain the substring 00.

- F_{n+2} is the number of subsets of the set $\{1, \ldots, n\}$ that contain no two consecutive integers.

Let's prove the second of these formulas. Let s_n be the number of binary strings of length n that contain no 00. We will prove that $s_n = F_{n+2}$, for $n \geq 1$. Observe that $s_1 = F_3 = 2$, and $s_2 = F_4 = 3$. We will show that $s_n = s_{n-1} + s_{n-2}$, for $n \geq 2$ (the same recurrence relation as the one satisfied by the Fibonacci numbers). Notice that each binary string of length n

that does not contain 00 ends in either 1 or 10. The number of such strings of the first type is s_{n-1} and the number of such strings of the second type is s_{n-2}. Hence $s_n = s_{n-1} + s_{n-2}$, for $n \geq 2$. Now, since $\{s_n\}$ satisfies the same recurrence relation as the Fibonacci numbers, and $s_1 = F_3$ and $s_2 = F_4$, it follows by mathematical induction that $s_n = F_{n+2}$, for all $n \geq 1$.

The next result, a very handy identity, was discovered by Giovanni Domenico Cassini (1625–1712).

Example 6.1. (Cassini's identity) Prove that $F_n^2 - F_{n-1}F_{n+1} = (-1)^{n+1}$, for $n \geq 1$.

Solution: We will prove the result by induction. The identity holds for $n = 1$, since $F_1^2 - F_0 F_2 = 1 - 0 = 1 = (-1)^2$. Assume that it holds for n. Then

$$F_{n+1}^2 - F_n F_{n+2} = F_{n+1}^2 - F_n(F_n + F_{n+1})$$
$$= F_{n+1}(F_{n+1} - F_n) - F_n^2$$
$$= F_{n+1}F_{n-1} - F_n^2$$
$$= (-1)^{n+2}.$$

Hence, the formula holds for $n + 1$ and by induction for all $n \geq 1$. ∎

Here is a wonderful (and perhaps little known) fact about Pascal's triangle.

Proposition 6.2 (David Singmaster, 1975). Infinitely many numbers occur at least six times in Pascal's triangle.

Proof. Consider solutions to

$$r = \binom{n}{m-1} = \binom{n-1}{m},$$

given by

$$m = F_{2k-1}F_{2k}, \quad n = F_{2k}F_{2k+1}, \quad k \geq 2.$$

The number r in such a solution occurs (at least) six times in Pascal's triangle:

$$\binom{r}{1} = \binom{r}{r-1} = \binom{n}{m-1} = \binom{n}{n-m+1} = \binom{n-1}{m} = \binom{n-1}{n-m-1}.$$

Check that these occurrences are really distinct!

Let's verify that such m and n give the claimed values of r. The following relations are equivalent:

$$\binom{n}{m-1} = \binom{n-1}{m}$$

$$\frac{n!}{(m-1)!(n-m+1)!} = \frac{(n-1)!}{m!(n-m-1)!}$$

$$mn = (n-m+1)(n-m)$$

$$F_{2k-1}F_{2k}F_{2k}F_{2k+1} = (F_{2k}F_{2k+1} - F_{2k-1}F_{2k} + 1)(F_{2k}F_{2k+1} - F_{2k-1}F_{2k})$$

$$= (F_{2k}(F_{2k+1} - F_{2k-1}) + 1)(F_{2k}(F_{2k+1} - F_{2k-1}))$$

$$= (F_{2k}^2 + 1)F_{2k}^2$$

$$F_{2k-1}F_{2k+1} = F_{2k}^2 + 1.$$

The final relation is true by Cassini's identity. ∎

The smallest such number given by our proof (when $k = 2$) is 3003.

Pascal's identity together with initial values (see p. 6) makes a recurrence formula that allows us to build Pascal's triangle:

$$\binom{n}{k} = \binom{n-1}{k-1} + \binom{n-1}{k}, \quad 1 \le k \le n,$$

$$\binom{n}{0} = \binom{n}{n} = 1, \quad n \ge 0.$$

We also have a direct formula for any given entry of Pascal's triangle:

$$\binom{n}{k} = \frac{n!}{k!(n-k)!}, \quad 0 \le k \le n.$$

Which is more useful, the recurrence formula or the direct formula? It depends on the situation.

A sequence $\{a_n\}$ satisfies a *linear homogeneous recurrence relation of order k with constant coefficients* if

$$a_n = \sum_{i=1}^{k} c_i a_{n-i},$$

for constants $c_1, \ldots, c_k$, and all $n \ge k$.

The Fibonacci sequence satisfies a linear homogeneous recurrence relation of order 2 with constant coefficients.

How do we find an explicit formula for the nth Fibonacci number? We will show how to guess and construct a solution. Assume that x^n, where $n \geq 0$, is the general term of a sequence that satisfies the Fibonacci recurrence relation (but not necessarily with the same initial values). Then

$$x^n = x^{n-1} + x^{n-2}.$$

Assuming that $x \neq 0$, we divide through by x^n and obtain the equation

$$x^2 - x - 1 = 0.$$

This polynomial $x^2 - x - 1$ is called the *characteristic polynomial* of the sequence. We use the quadratic formula to find the two roots of the characteristic polynomial:

$$\phi = \frac{1 + \sqrt{5}}{2}, \quad \hat{\phi} = \frac{1 - \sqrt{5}}{2}.$$

We call ϕ the "golden ratio." Note that $\phi \doteq 1.6$ and $\hat{\phi} \doteq -0.6$.

So we know that ϕ^n and $\hat{\phi}^n$ both satisfy the Fibonacci recurrence relation. Any linear combination of the basic solutions, $A\phi^n + B\hat{\phi}^n$, with $A, B \in \mathbf{R}$, also satisfies the recurrence relation, for

$$(A\phi^{n-1} + B\hat{\phi}^{n-1}) + (A\phi^{n-2} + B\hat{\phi}^{n-2}) = A(\phi^{n-1} + \phi^{n-2}) + B(\hat{\phi}^{n-1} + \hat{\phi}^{n-2})$$

$$= A\phi^n + B\hat{\phi}^n.$$

We use the initial values to solve for the coefficients A and B. Recalling that $F_0 = 1$ and $F_1 = 1$, we obtain two linear equations to solve simultaneously:

$$1 = A\phi^0 + B\hat{\phi}^0 = A + B$$

$$1 = A\phi^1 + B\hat{\phi}^1 = A\left(\frac{1 + \sqrt{5}}{2}\right) + B\left(\frac{1 - \sqrt{5}}{2}\right).$$

We find that

$$A = \frac{1}{\sqrt{5}} \text{ and } B = -\frac{1}{\sqrt{5}}.$$

Thus, the general formula for the Fibonacci numbers is

$$F_n = \frac{1}{\sqrt{5}}\phi^n - \frac{1}{\sqrt{5}}\hat{\phi}^n, \quad n \geq 0.$$

The above function satisfies the recurrence relation and initial values of the Fibonacci sequence, and hence is a formula for the Fibonacci sequence (since the sequence is well-defined).

What is the growth rate of $\{F_n\}$? We say that a positive-valued function $f(n)$ is "asymptotic" to another such function $g(n)$, and we write $f(n) \sim g(n)$, if $\lim_{n\to\infty} f(n)/g(n) = 1$. Since $\hat{\phi}^n \to 0$ as $n \to \infty$, we conclude that

$$F_n \sim \frac{\phi^n}{\sqrt{5}}.$$

If f and g are functions defined on the set of positive integers, and

$$|f(n)| \le C|g(n)|$$

for some positive constant C and all $n \ge n_0$, for some n_0, then we say that f is "big oh" of g, and write $f(n) = O(g(n))$. In this notation,

$$F_n = O\left(\phi^n\right).$$

Example 6.3. Find an explicit formula for the sequence $\{a_n\}$ defined by the recurrence formula

$$a_0 = 1, \ a_1 = 1, \ a_n = 6a_{n-1} - 9a_{n-2}, \quad n \ge 2.$$

Solution: The characteristic polynomial of the sequence is

$$x^2 - 6x + 9 = (x - 3)^2,$$

which has 3 as a double root. Hence, 3^n is a solution to the recurrence relation. However, we need a second solution in order to make the formula satisfy the initial values. A guess for a second solution is $n3^n$. Let's check that this solution satisfies the recurrence relation:

$$6(n - 1)3^{n-1} - 9(n - 2)3^{n-2} = 3^{n-2}(18n - 18 - 9n + 18)$$

$$= n3^n.$$

Any linear combination of our two solutions also satisfies the recurrence relation:

$$A3^n + Bn3^n.$$

In order to satisfy the initial values, $a_0 = 1$ and $a_1 = 1$, we require that

$$1 = A$$

$$1 = 3A + 3B,$$

and hence $A = 1$ and $B = -2/3$. Therefore, the explicit formula for the sequence is

$$a_n = 3^n - 2n3^{n-1}, \quad n \ge 0.$$

We see that $a_n = O(n3^n)$. ∎

The next example illustrates the technique of adding a particular solution and a homogeneous solution.

Example 6.4. Find an explicit formula for the sequence $\{a_n\}$ defined by the recurrence formula

$$a_0 = 1, \; a_1 = 1, \; a_n = 6a_{n-1} - 9a_{n-2} + n, \quad n \geq 2.$$

Solution: We find a particular solution to the recurrence relation. Assume the existence of a solution of the form $a_n = \alpha n + \beta$. Thus

$$\alpha n + \beta = 6(\alpha(n-1) + \beta) - 9(\alpha(n-2) + \beta) + n$$

$$(4\alpha - 1)n = 12\alpha - 4\beta.$$

In order for this identity to hold for all n, we must have $\alpha = 1/4$ and hence $\beta = 3/4$. Therefore

$$\frac{1}{4}n + \frac{3}{4}$$

satisfies the recurrence relation.

We solved the homogeneous version of this recurrence relation in the previous example. Thus, the general solution to the recurrence relation is of the form

$$A3^n + Bn3^n + \frac{1}{4}n + \frac{3}{4}.$$

The initial values, $a_0 = 1$ and $a_1 = 1$, determine the values $A = 1/4$ and $B = -1/4$. Therefore, an explicit formula is

$$a_n = \frac{1}{4}3^n - \frac{1}{4}n3^n + \frac{1}{4}n + \frac{3}{4}, \quad n \geq 0.$$

∎

The *Lucas numbers*, named after François Édouard Anatole Lucas (1842–1891), are defined as

$$L_0 = 2, \; L_1 = 1, \; L_n = L_{n-1} + L_{n-2}, \quad n \geq 2.$$

Thus, the Lucas numbers are

2, 1, 3, 4, 7, 11, 18, 29, 47, 76, 123, 200, 323, 523, 846, 1369, ...

- L_n is the number of ways an $n \times 1$ box may be packed with 2×1 and 1×1 boxes, allowing "wrap-around."

Since the Lucas numbers satisfy the same recurrence relation as the Fibonacci numbers, they have the same characteristic polynomial, $x^2 - x - 1$. Taking into account the initial values $L_0 = 2$ and $L_1 = 1$, we obtain a formula for the Lucas numbers:

$$L_n = \phi^n + \hat{\phi}^n, \quad n \geq 0.$$

The simplicity of this formula is a shining property of the Lucas sequence.

Exercises

1. Prove the identity
$$F_1 + \cdots + F_n = F_{n+2} - 1, \quad n \geq 1.$$

2. Prove the identity
$$F_1^2 + \cdots + F_n^2 = F_n F_{n+1}, \quad n \geq 1.$$

3. Prove the identity
$$F_{m+n} = F_m F_{n+1} + F_{m-1} F_n, \quad m \geq 1, n \geq 0.$$

◇4. Use a computer and the Fibonacci recurrence formula to calculate F_{100}.

5. Where do you find Fibonacci numbers in Pascal's triangle? What identity supports this?

6. Find positive integers n, k, with $k < n$, for which
$$\binom{n}{k} + \binom{n}{k+1} = \binom{n}{k+2}.$$

⋆7. Prove that
$$\sum_{n=1}^{\infty} \tan^{-1} \frac{1}{F_{2n+1}} = \frac{\pi}{4}.$$

◇8. Use a computer to find the smallest number other than 1 that appears six times in Pascal's triangle.

◇9. Use a computer to find the second smallest number given by Proposition 6.2 that appears six times in Pascal's triangle.

10. Let $\{a_n\}$ be defined by the recurrence
$$a_0 = 0, \ a_1 = 1, \ a_n = 5a_{n-1} - 6a_{n-2}, \ n \geq 2.$$
Find an explicit formula for a_n.

11. Suppose that the sequence $\{a_n\}$ satisfies the recurrence relation
$$a_n = 3a_{n-1} + 4a_{n-2} - 12a_{n-3}, \quad n \geq 3,$$
where $a_0 = 0$, $a_1 = 1$, and $a_2 = 2$. Find an explicit formula for a_n.

12. Let $\{b_n\}$ be defined by the recurrence

$$b_0 = 0,\ b_1 = 0,\ b_2 = 1,\ b_n = 4b_{n-1} - b_{n-2} - 6b_{n-3},\ n \geq 3.$$

Find an explicit formula for b_n.

Do the same where the initial values are $b_0 = 0,\ b_1 = 1,\ b_2 = 2$.

13. Define $\{a_n\}$ by the recurrence

$$a_0 = 0,\ a_1 = 1,\ a_n = 5a_{n-1} - 6a_{n-2},\ n \geq 2$$

and $\{b_n\}$ by the recurrence

$$b_0 = 0,\ b_1 = 1,\ b_n = 9b_{n-1} - 20b_{n-2},\ n \geq 2.$$

Find a linear recurrence for the sequence $\{c_n\}$ defined by

$$c_n = a_n + b_n, \quad n \geq 0.$$

Find a linear recurrence for the sequence $\{d_n\}$ defined by

$$d_n = a_n b_n, \quad n \geq 0.$$

14. Find a linear recurrence for the sequence $\{a_n\}$ defined by

$$a_n = F_n + 2^n, \quad n \geq 0,$$

where F_n is the nth Fibonacci number.

15. (a) Find a linear homogeneous recurrence formula for the sequence $\{a_n\}$ defined by $a_n = 3^n + n^2$, where $n \geq 0$.

(b) Find a linear homogeneous recurrence formula for the sequence $\{a_n\}$ defined by $a_n = 3^n + n^2 + 6n + 7$, where $n \geq 0$.

16. Find an explicit formula for the sequence $\{a_n\}$ defined by the recurrence formula

$$a_0 = 0,\ a_1 = 1,\ a_n = a_{n-1} + a_{n-2} + n,\ n \geq 2.$$

17. Find an explicit formula for the sequence $\{a_n\}$ defined by the recurrence formula

$$a_0 = 0,\ a_1 = 1,\ a_n = a_{n-1} + a_{n-2} + 2^n,\ n \geq 2.$$

18. Prove the identity $L_n = F_{n-1} + F_{n+1}$, for $n \geq 1$.

19. Prove the identity $F_n = (L_{n-1} + L_{n+1})/5$, for $n \geq 1$.

20. Prove the identity $F_{2n} = F_n L_n$, for $n \geq 0$.

21. Find a relation for Lucas numbers similar to Cassini's identity.

22. Find a linear recurrence relation satisfied by all cubic polynomials.

23. A *square number* is an integer of the form n^2. A *triangular number* is an integer of the form $1 + 2 + \cdots + n = n(n+1)/2$. Let a_n be the nth number that is both square and triangular. For example, $a_0 = 0$, $a_1 = 1$, and $a_2 = 36$. Find a linear homogeneous recurrence relation with constant coefficients for $\{a_n\}$.

24. Find a linear recurrence relation with constant coefficients for the sequence $\{2^n F_n\}$.

25. Find a linear recurrence relation with constant coefficients for the sequence of squares of the Fibonacci numbers, $\{F_n^2\}$.

⋆26. Prove that the kth powers of the Fibonacci numbers satisfy a linear homogeneous recurrence relation of order $k + 1$ with integer coefficients.

Part III

Counting: Advanced

Chapter 7

Generating Functions and Making Change

There are 293 ways to make change for a dollar.

There are 88,265,881,340,710,786,348,934,950,201,250,975,072,332,541,120,001 ways to make change for a million dollars.

Given any sequence $a_0, a_1, a_2, \ldots$, we define the *(ordinary) generating function*

$$f(x) = \sum_{n=0}^{\infty} a_n x^n = a_0 + a_1 x + a_2 x^2 + a_3 x^3 + \cdots.$$

The infinite series may or may not converge.

Example 7.1. Find the ordinary generating function for the Fibonacci sequence $\{F_0, F_1, F_2, \ldots\}$.

Solution: Let $f(x) = \sum_{n=0}^{\infty} F_n x^n$. Then

$$
\begin{aligned}
f(x) &= x + x^2 + 2x^3 + 3x^4 + 5x^5 + \cdots \\
xf(x) &= x^2 + x^3 + 2x^4 + 3x^5 + 5x^6 + \cdots \\
x^2 f(x) &= x^3 + x^4 + 2x^5 + 3x^6 + 5x^7 + \cdots.
\end{aligned}
$$

Through mass-cancellation, the recurrence relation for the Fibonacci numbers yields

$$f(x) - xf(x) - x^2 f(x) = x$$

and hence

$$f(x) = \frac{x}{1 - x - x^2}.$$

■

The generating function for the Fibonacci sequence can be used to find the direct formula for F_n that we found in Chapter 6.

Notice that the generating function for the Fibonacci sequence is a rational function. In general, a sequence satisfies a linear homogeneous recurrence relation with constant coefficients if and only if it has a rational ordinary generating function of a certain type. We will prove this shortly.

Also, notice that the denominator of the generating function for the Fibonacci numbers, $1 - x - x^2$, takes its form from the Fibonacci recurrence, while the numerator comes from multiplying the generating function by the denominator and keeping only those terms of degree less than 2.

Recall that a sequence $\{a_n\}$ satisfies a linear homogeneous recurrence relation of order k with constant coefficients $c_1, \ldots, c_k$ if

$$a_n = \sum_{i=1}^{k} c_i a_{n-i},$$

for all $n \geq k$.

Theorem 7.2. Given a sequence $\{a_n\}$ and arbitrary numbers $c_1, \ldots, c_k$, the following three assertions are equivalent.

(1) The sequence $\{a_n\}$ satisfies a linear recurrence relation with constant coefficients $c_1, \ldots, c_k$, i.e.,

$$a_n = \sum_{i=1}^{k} c_i a_{n-i},$$

for $n \geq k$.

(2) The sequence $\{a_n\}$ has a rational ordinary generating function of the form

$$\frac{g(x)}{1 - \sum_{i=1}^{k} c_i x^i},$$

where g is a polynomial of degree at most $k - 1$.

(3) If

$$1 - \sum_{i=1}^{k} c_i x^i = (1 - r_1 x)(1 - r_2 x) \ldots (1 - r_k x),$$

with the r_i distinct, then

$$a_n = \alpha_1 r_1^n + \cdots + \alpha_k r_k^n,$$

for all $n \geq 0$, and constants $\alpha_1, \ldots, \alpha_k$.

More generally, if

$$1 - \sum_{i=1}^{k} c_i x^i = (1 - r_1 x)^{m_1} (1 - r_2 x)^{m_2} \ldots (1 - r_l x)^{m_l},$$

where the roots $r_1, \ldots, r_l$ occur with multiplicities $m_1, \ldots, m_l$, then

$$a_n = p_1(n) r_1^n + \cdots + p_l(n) r_l^n$$

for all $n \geq 0$ and polynomials $p_1, \ldots, p_l$, where $\deg p_j < m_j$ for $1 \leq j \leq l$.

The proof proceeds along the lines of Example 7.1, although the case of repeated roots of the characteristic polynomial requires partial fractions decompositions.

Note. The factorization of $1 - \sum_{i=1}^{k} c_i x^i$ called for in the proof (and in practice) can be accomplished using the change of variables $y = 1/x$. Then

$$1 - \sum_{i=1}^{k} c_i x^i = 1 - \sum_{i=1}^{k} c_i y^{-i} = y^{-k} \left(y^k - \sum_{i=1}^{k} c_i y^{k-i} \right).$$

The problem is reduced to factoring the polynomial

$$y^k - \sum_{i=1}^{k} c_i y^{k-i}.$$

This polynomial is the characteristic polynomial of the recurrence relation.

Example 7.3. Find the generating function for the sequence defined by the recurrence relation $a_n = 6a_{n-1} - 9a_{n-2}$, for $n \geq 2$, and $a_0 = 1$, $a_1 = 1$. (This comes from Example 6.3.) Use the generating function to find a direct formula for a_n.

Solution: The form of the recurrence relation tells us that the denominator of the generating function is $1 - 6x + 9x^2$. To get the numerator, we calculate $(1 - 6x + 9x^2)(a_0 + a_1 x) = (1 - 6x + 9x^2)(1 + x) = 1 - 5x + \cdots$. The only terms of degree less than 2 are $1 - 5x$, so the numerator is $1 - 5x$. Hence, the generating function is

$$\frac{1 - 5x}{1 - 6x + 9x^2}.$$

To find a direct formula for a_n, we write the generating function as

$$(1 - 5x)(1 - 3x)^{-2}.$$

Thus, we have a binomial series with a negative exponent. The expansion is

$$(1 - 5x) \sum_{k=0}^{\infty} (-1)^k 3^k \binom{-2}{k} x^k.$$

Therefore

$$a_n = (-1)^n 3^n \binom{-2}{n} - 5(-1)^{n-1} 3^{n-1} \binom{-2}{n-1}$$

$$= 3^n \binom{n+1}{n} - 5 \cdot 3^{n-1} \binom{n}{n-1}$$

$$= 3^n (n+1) - 5n3^{n-1}$$

$$= 3^n - 2n3^{n-1}, \quad n \geq 0.$$

This is the same solution that we saw before. ∎

If $f(x)$ is the ordinary generating function for a sequence $\{a_n\}$, then

$$a_n = \frac{f^{(n)}(0)}{n!}, \quad n \geq 0.$$

Similarly, if $f(x,y)$ is the ordinary generating function (in two variables) for a sequence $\{a_{m,n}\}$, then

$$a_{m,n} = \frac{\partial_x^m \partial_y^n f(0,0)}{m!n!}, \quad m \geq 0, n \geq 0.$$

Example 7.4. How many ways can you make change for $1.00, using units of 0.01, 0.05, 0.10, 0.25, 0.50, and 1.00? Here are some examples:

$$5 + 10 + 10 + 25 + 50$$

$$1 + 1 + 1 + 1 + 1 + 1 + 1 + 1 + 1 + 1 + 10 + 10 + 10 + 25 + 25$$

$$25 + 25 + 25 + 25.$$

(We write the summands in terms of cents, with no decimals.)

Solution: We solve the problem by generalizing. For $n \geq 0$, let a_n be the number of ways to make change for an amount n. For convenience, we set $a_0 = 1$. It's easy to work out the first few values of the sequence $\{a_n\}$, so we see that its generating function looks like

$$1 + 1x + 1x^2 + 1x^3 + 1x^4 + 2x^5 + 2x^6 + 2x^7 + 2x^8 + 2x^9 + 4x^{10} + \cdots.$$

We claim that this generating function is the rational function

$$\frac{1}{(1-x)(1-x^5)(1-x^{10})(1-x^{25})(1-x^{50})(1-x^{100})}.$$

Using a computer algebra system, one finds that the coefficient of x^{100} of this generating function is 293, i.e., there are 293 ways to make change for a dollar. Essentially, the computation amounts to finding the coefficient of x^{100} in the generating function product

$$(1 + x + \cdots + x^{100})$$

$$\cdot (1 + x^5 + \cdots + x^{100})$$

$$\cdot (1 + x^{10} + \cdots + x^{100})$$

$$\cdot (1 + x^{25} + \cdots + x^{100})$$

$$\cdot (1 + x^{50} + \cdots + x^{100})$$

$$\cdot (1 + x^{100}).$$

In order to explain the generating function, observe that the factors in the denominator give rise to geometric series. For example, the second factor gives

$$\frac{1}{(1 - x^5)} = 1 + x^5 + x^{2 \cdot 5} + x^{3 \cdot 5} + x^{4 \cdot 5} + x^{5 \cdot 5} + x^{6 \cdot 5} + \cdots.$$

In the product, each term corresponds to a way to make change for a dollar. For instance, the term corresponding to the sum $5 + 10 + 10 + 25 + 50$ is shown in boldface:

$$(\mathbf{1} + x + x^2 + x^3 + x^4 + x^5 + x^6 + \cdots)$$

$$\cdot (1 + \mathbf{x^5} + x^{2 \cdot 5} + x^{3 \cdot 5} + x^{4 \cdot 5} + x^{5 \cdot 5} + x^{6 \cdot 5} + \cdots)$$

$$\cdot (1 + x^{10} + \mathbf{x^{2 \cdot 10}} + x^{3 \cdot 10} + x^{4 \cdot 10} + x^{5 \cdot 10} + x^{6 \cdot 10} + \cdots)$$

$$\cdot (1 + \mathbf{x^{25}} + x^{2 \cdot 25} + x^{3 \cdot 25} + x^{4 \cdot 25} + x^{5 \cdot 25} + x^{6 \cdot 25} + \cdots)$$

$$\cdot (1 + \mathbf{x^{50}} + x^{2 \cdot 50} + x^{3 \cdot 50} + x^{4 \cdot 50} + x^{5 \cdot 50} + x^{6 \cdot 50} + \cdots)$$

$$\cdot (1 + x^{100} + x^{2 \cdot 100} + x^{3 \cdot 100} + x^{4 \cdot 100} + x^{5 \cdot 100} + x^{6 \cdot 100} + \cdots).$$

By the way, since the denominator of the generating function is a polynomial of order 191, the sequence $\{a_n\}$ satisfies a linear recurrence relation of order 191. ∎

Example 7.5. How many ways can you make $1 million using any number of pennies, nickels, dimes, quarters, half-dollars, one-dollar bills, five-dollar bills, ten-dollar bills, twenty-dollar bills, fifty-dollar bills, and hundred-dollar bills?

Solution: The generating function that counts the number of ways to make change using the given denominations is

$$f(x) = \frac{1}{1-x}\frac{1}{1-x^5}\frac{1}{1-x^{10}}\frac{1}{1-x^{25}}\frac{1}{1-x^{50}}\frac{1}{1-x^{100}}$$

$$\cdot \frac{1}{1-x^{500}}\frac{1}{1-x^{1000}}\frac{1}{1-x^{2000}}\frac{1}{1-x^{5000}}\frac{1}{1-x^{10000}}.$$

Our job is to find $a_{100,000,000}$, the coefficient of $x^{100,000,000}$ in this generating function. The idea is to manipulate the generating function to make the task easier.

Every exponent of x in the denominator is a multiple of 5 except for in the first factor. Hence, we rewrite the factor $(1-x)^{-1}$ as

$$(1+x+x^2+x^3+x^4)/(1-x^5),$$

and define

$$\hat{f}(x) = \frac{1}{1-x}\frac{1}{1-x}\frac{1}{1-x^2}\frac{1}{1-x^5}\frac{1}{1-x^{10}}\frac{1}{1-x^{20}}$$

$$\cdot \frac{1}{1-x^{100}}\frac{1}{1-x^{200}}\frac{1}{1-x^{400}}\frac{1}{1-x^{1000}}\frac{1}{1-x^{2000}},$$

so that

$$f(x) = (1+x+x^2+x^3+x^4)\hat{f}(x^5).$$

Since \$1 million is a multiple of 5 (cents), the terms x, x^2, x^3, and x^4 in the first factor don't matter. In the denominator of $\hat{f}(x)$, all the powers of x divide the largest power, 2000. Accordingly, we rewrite each factor in the denominator as $(1-x^{2000})$, with a compensating factor in the numerator. The new numerator is

$$(1+x+\cdots+x^{1999})^2(1+x^2+x^4+\cdots+x^{1998})$$

$$\cdot(1+x^5+\cdots+x^{1995})(1+x^{10}+\cdots+x^{1990})(1+x^{20}+\cdots+x^{1980})$$

$$\cdot(1+x^{100}+\cdots+x^{1900})(1+x^{200}+\cdots+x^{1800})$$

$$\cdot(1+x^{400}+\cdots+x^{1600})(1+x^{1000}).$$

A computer algebra system can multiply out the new numerator in a moment. Now the denominator looks like $(1-x^{2000})^{11}$, which we expand as a binomial series:

$$(1-x^{2000})^{-11} = \sum_{k=0}^{\infty}\binom{k+10}{10}x^{2000k}.$$

We complete our calculation by multiplying the appropriate terms from the numerator and this binomial series to obtain $\hat{a}_{20,000,000}$, the coefficient of $x^{20,000,000}$ in $\hat{f}$. The numerator is a polynomial of degree 18261, but the only powers of x that matter are multiples of 2000; the corresponding coefficients are, say,

$$\alpha_0 = 1$$

$$\alpha_{2000} = 1424039612928$$

$$\alpha_{4000} = 212561825179035$$

$$\alpha_{6000} = 3224717280609587$$

$$\alpha_{8000} = 11601166434205649$$

$$\alpha_{10000} = 12519790995056639$$

$$\alpha_{12000} = 4102067385934937$$

$$\alpha_{14000} = 334900882733305$$

$$\alpha_{16000} = 3371148659578.$$

$$\alpha_{18000} = 8008341.$$

Finally, we calculate

$$a_{100,000,000} = \hat{a}_{20,000,000} = \sum_{j=0}^{9} \alpha_{2000j} \binom{10000 + 10 - j}{10}$$

$$= 88265881340710786348934950201250975072332541120001$$

$$\doteq 8.8 \times 10^{49}.$$

■

Exercises

1. Evaluate the infinite sum $\sum_{n=1}^{\infty} nF_n/3^n$.

2. Find the ordinary generating function for the Lucas numbers.

3. Find the ordinary generating function for the sequence $\{a_n\}$ given by the recurrence formula

$$a_0 = 0, \ a_1 = 1, \ a_n = 5a_{n-1} - 6a_{n-2}, \ n \geq 2.$$

4. Evaluate the infinite series

$$\sum_{n=0}^{\infty} \frac{(-1)^n a_n}{10^n},$$

where $\{a_n\}$ is the sequence of the previous exercise.

5. Find a recurrence formula for the coefficient of x^n in the series expansion of $(1 + 3x + x^2)^{-1}$.

6. Let a_n be the coefficient in the previous exercise. Prove that $a_n = (-1)^n F_{2n+2}$.

7. Find a recurrence formula for the coefficient of x^n in the series expansion of $(1 + x + 2x^2)^{-1}$.

8. Find the generating function for the sequence of perfect squares, $\{n^2\}$, for $n \geq 0$.

◇9. When Professor Bumble left a tip at a restaurant, he noticed that the amount he left, n, can be given in n different ways using the units 1, 5, 10, 25, 50, and 100. What is n?

◇10. Use a computer and an appropriate generating function to determine the number of ways of making change for \$1 using an even number of coins.

11. Suppose that the units of money are 1, 5, 10, 25, 50, and 100. Show that for every positive integer n, there are more ways to make n using an even number of these coins than using an odd number if n is even, and more ways to make n using an odd number of these coins than using an even number if n is odd. Show that the same result holds for any system of coins $\mathcal{S}$ with the property that $2k \in \mathcal{S} \implies k \in \mathcal{S}$.

12. Show that we can calculate the number of ways to make change for a dollar in the following way. Let P_n be the number of ways to make change for an amount n, given that the highest denomination used is a penny. Similarly, define N_n, D_n, Q_n, H_n, and W_n to be the number of ways to make change for n given that the highest denomination used is a nickel, dime, quarter, half-dollar, and whole dollar, respectively.

Explain why the following relations are true.

$$P_n = P_{n-1}$$

$$N_n = P_{n-5} + N_{n-5}$$

$$D_n = P_{n-10} + N_{n-10} + D_{n-10}$$

$$Q_n = P_{n-25} + N_{n-25} + D_{n-25} + Q_{n-25}$$

$$H_n = P_{n-50} + N_{n-50} + D_{n-50} + Q_{n-50} + H_{n-50}$$

$$W_n = P_{n-100} + N_{n-100} + D_{n-100} + Q_{n-100} + H_{n-100} + W_{n-100}$$

Use these relations to find the number of ways to make change for a dollar.

13. (a) Show that the generating function (in two variables) for binomial coefficients is

$$\frac{1}{1 - x - y}.$$

(b) Show that the generating function (in three variables) for multinomial coefficients of the form $\binom{n}{n_1,n_2,n_3}$ is

$$\frac{1}{1 - x - y - z}.$$

◇14. Use a generating function to determine the number of solutions in nonnegative integers to the equation

$$a + 2b + 4c = 10^{30}.$$

◇15. Determine the number of solutions in nonnegative integers to the equation

$$a + 2b + 3c = 10^{30}.$$

◇16. Determine the number of solutions in nonnegative integers to the equation

$$a + b + 4c = 10^{30}.$$

⋆17. Prove that, for each positive integer k, there exists a monic polynomial $p(n)$ of degree $k + 1$ with integer coefficients such that

$$\sum_{i=1}^{n} i^k \binom{n}{i} = 2^{n-k} p(n).$$

18. Find a linear homogeneous recurrence relation (not with constant coefficients) for the sequence $\{a_n\}$, where $a_n = 2^n + n!$.

★19. (Frobenius' stamp problem) Let a and b be integers greater than 1 with no common factor. Given an unlimited supply of stamps in the denominations a and b, show that the set of positive integer amounts that cannot be made with these stamps is finite and find the size of the set. Find a formula for the largest value that cannot be made.

   The problem is due to Ferdinand Georg Frobenius (1849–1917).

◇20. How many ways can you make $1 million using any number of pennies, nickels, dimes, quarters, half-dollars, and one-dollar bills?

◇★21. How many ways can you make $1 million using any number of pennies, nickels, dimes, quarters, half-dollars, one-dollar bills, two-dollar bills, five-dollar bills, ten-dollar bills, twenty-dollar bills, fifty-dollar bills, and hundred-dollar bills?

Chapter 8

Integer Triangles

The number of incongruent triangles with integer side lengths and perimeter 10^{100} is

$$208\underbrace{3\ldots3}_{196}.$$

How do we arrive at such a number? Let's first solve some simpler problems. How many incongruent triangles have integer side lengths and perimeter 10? There are only two: $(2, 4, 4)$ and $(3, 3, 4)$. (We specify a triangle by giving the ordered triple of its side lengths in nondecreasing order. A triple (a, b, c) must satisfy the triangle inequality $a + b > c$.)

Let $t(n)$ be the number of integer triangles of perimeter n. Let's generate some data. It is convenient to set $t(0) = 0$.

n	0	1	2	3	4	5	6	7	8
$t(n)$	0	0	0	1	0	1	1	2	1

The sequence $\{t(n)\}$ is known as Alcuin's sequence, after Alcuin of York (735–804).

If we generate more data and plot the values of $t(n)$, we are led to the conjecture that the function is nearly a quadratic polynomial of the form $n^2/48$. We can also guess this from a rough estimate. For three side lengths to satisfy the triangle inequality, it is necessary and sufficient that the sum of any two of them is less than $n/2$. So, there are about $n/2$ choices for, say, x, and given x, the value of $x + y$ must be between $n/2$ and $n/2 + x$. This defines z as $z = n - x - y$. Hence, the number of choices of x, y, and z is about

$$\sum_{x=1}^{n/2} x \sim \frac{(n/2)(n/2+1)}{2} \sim \frac{n^2}{8}.$$

Most of the time, x, y, and z will be different, so the ordered triples (a, b, c) have been over-counted by a factor of $3! = 6$. Therefore $t(n)$ is approximately $n^2/48$. A little tinkering yields the formula

$$t(n) = \begin{cases} \left\{\frac{n^2}{48}\right\} & \text{if } n \text{ is even,} \\ \left\{\frac{(n+3)^2}{48}\right\} & \text{if } n \text{ is odd,} \end{cases}$$

where $\{x\}$ is the nearest integer to x.

We need to prove the formula, but if it's true then it's a simple matter to compute the number of incongruent triangles with integer sides and perimeter 10^{100}. We have $t(10^{100}) = 208\underbrace{3\ldots3}_{196}$ (by direct calculation), as claimed in the teaser at the beginning of this chapter.

Note. We should show that our formula for $t(n)$ is well-defined, that is, $\{x^2/48\}$ cannot be half-way between two integers. If $x^2/48 = i + 1/2$, where x and i are integers, then $x^2 = 48i + 24$. Since 8 divides the right side of this relation, 8 must divide the left side. The left side is a perfect square, so it is divisible by 16, but the right side is not divisible by 16 (the first term is and the second term isn't). This is a contradiction.

Let's prove our formula. We claim that the generating function

$$t(0) + t(1)x + t(2)x^2 + t(3)x^3 + \cdots$$

is the rational function

$$\frac{x^3}{(1-x^2)(1-x^3)(1-x^4)}.$$

The idea is that we can "build up" to any given triangle (a, b, c) starting with the triangle $(1, 1, 1)$. The key observation is that we can write

$$(a, b, c) = (1, 1, 1) + \alpha(0, 1, 1) + \beta(1, 1, 1) + \gamma(1, 1, 2),$$

where α, β, and γ are determined uniquely (just solve for them, given $a + b + c = n$). The vectors $(0, 1, 1)$, $(1, 1, 1)$, and $(1, 1, 2)$ satisfy the weak triangle inequality $a + b \geq c$, and this is sufficient since we start with a non-degenerate triangle. As $2\alpha + 3\beta + 4\gamma = a + b + c - 3 = n - 3$, this shows that $t(n)$ is equal to the number of partitions of $n - 3$ where the parts are 2s, 3s, and 4s (order of terms unimportant). That is precisely what the generating function generates.

We can determine the approximate value $t(n) \sim n^2/48$ from the generating function by noting that the zeros of the denominator polynomial are all on the unit circle in the complex plane, with the zero of largest order being $z = 1$ (of order 3). By the binomial series expansion,

$$t(n) \sim \frac{C}{2!}n^2,$$

where

$$C = \lim_{z \to 1} \left[(1-z)^3 \frac{z^3}{(1-z^2)(1-z^3)(1-z^4)} \right] = \frac{1}{24}.$$

The denominator of the rational generating function yields a recurrence formula for the sequence $\{t(n)\}$ (expand the polynomial and look at its form), namely,

$$t(n) = t(n-2)+t(n-3)+t(n-4)-t(n-5)-t(n-6)-t(n-7)+t(n-9), \quad n \geq 9.$$

The initial values of the sequence are given in our table above.

Finally, we can use the recurrence relation to prove our formula for $t(n)$ by induction. For the moment, let's call the proposed formula $\hat{t}(n)$. We want to show that $\hat{t}(n) = t(n)$, for $n \geq 0$. By direct calculation, we see that

$$\hat{t}(n+24) = \hat{t}(n) + n + 12, \quad \text{if } n \text{ is even}$$

$$\hat{t}(n+24) = \hat{t}(n) + n + 15, \quad \text{if } n \text{ is odd}.$$

It's easy to use our known recurrence relation to calculate $t(n)$ for $0 \leq n \leq 32$ and find that $\hat{t}(n) = t(n)$ for these values.

For the induction hypothesis, assume that $\hat{t}(n) = t(n)$ for a "block" of 24 consecutive integers. Then, rewriting our recurrence relation, we have

$$\hat{t}(n)+\hat{t}(n-5)+\hat{t}(n-6)+\hat{t}(n-7) = \hat{t}(n-2)+\hat{t}(n-3)+\hat{t}(n-4)+\hat{t}(n-9)$$

for n in this block. Notice that whether n is even or odd, there are two even values of the argument and two odd values of the argument on the left side and the same on the right side. Thus, adding two terms of $n+12$ and two terms of $n+15$ on each side and $-5-6-7$ on the left and $-2-3-4-9$ on the right, we obtain

$$\hat{t}(n+24) + \hat{t}(n+24-5) + \hat{t}(n+24-6) + \hat{t}(n+24-7)$$

$$= \hat{t}(n+24-2) + \hat{t}(n+24-3) + \hat{t}(n+24-4) + \hat{t}(n+24-9).$$

Hence, $\hat{t}(n) = t(n)$ for the next block of 24 consecutive integers. By induction, $\hat{t}(n) = t(n)$ for all $n \geq 0$.

While not as famous as the Fibonacci sequence, Alcuin's sequence has some marvelous properties (as we will see in the exercises), and might repay further study.

Exercises

1. Find the number of incongruent triangles with integer sides and perimeter 1000.

$\diamond$2. Derive the formula for $t(n)$ using the generating function and partial fractions.

3. Find some pairs of (non-similar) integer triangles that have a common angle.

$\diamond$4. Find the smallest integer triangle for which the measure of one angle is twice that of another angle.

5. Find infinitely many (non-similar) integer triangles with one angle equal to 60°.

$\diamond$6. Let R and r represent the circumradius and inradius of a triangle, respectively. Find an integer triangle for which $R/r = 26$. Investigate which ratios R/r are possible for integer triangles.

7. Find a formula for the number of scalene integer triangles of perimeter n. What is the generating function for the number of such triangles?

8. Prove that Alcuin's sequence $\{t(n)\}$ is a zigzag sequence (its values alternately rise and fall), for $n \geq 6$.

9. Professor Bumble announces that there is exactly one integer n for which there are exactly n integer triangles with perimeter n. What is n?

10. Prove that Alcuin's sequence $\{t(n)\}$ satisfies the recurrence relation

$$t(n) = 3t(n-12) - 3t(n-24) + t(n-36), \quad n \geq 36.$$

$\star$11. Consider Alcuin's sequence modulo 2. We obtain a repeating cycle of length 24:

$$0, 0, 0, 1, 0, 1, 1, 0, 1, 1, 0, 0, 1, 1, 0, 1, 1, 0, 1, 0, 0, 0, 0, 0.$$

We say that the period mod 2 is 24. Prove that, given any modulus $m \geq 2$, the period of Alcuin's sequence mod m is $12m$.

12. Find infinitely many (non-similar) pairs of integer triangles such that the two triangles in each pair have the same perimeter and the same area.

Chapter 9

Rook Paths and Queen Paths

A chess Rook may travel in 470,010 ways from one corner square of a chessboard to the opposite corner square. A Queen may make the same trip in 1,499,858 ways.

A chess Rook may move any number of squares horizontally or vertically in one step. How many paths can a chess Rook take from the lower-left corner square to the upper-right corner square of an ordinary 8×8 chessboard? Assume that the Rook moves right or up at each step. An example of such a Rook path is shown in Figure 9.1.

Notice that the pausing points are important. Without them, the number of paths is easily counted by the binomial coefficient $\binom{14}{7}$. This is the famous problem of counting paths along city blocks from one intersection to another (as on p. 13).

It's often helpful to generalize a problem. Rook paths to any given square are equivalent to lattice paths that start at $(0,0)$ and move by steps of the form $(x,0)$ or $(0,y)$, where x and y are positive integers, toward a goal point (m,n), where $m, n \geq 0$. Let's call the number of paths $a(m,n)$, where $m, n \geq 0$. Starting with the value $a(0,0) = 1$, we calculate each other number in turn by adding all the entries to the left of that number and below that number. For example, $a(3,2) = 2 + 5 + 14 + 4 + 12 = 37$. The reason for this rule is that the Rook must arrive at that particular square from one of the squares to its left or below it. Thus, the number of Rook paths from the lower-left corner to the upper-right corner of the chessboard is $a(7,7) = 470010$.

$\vdots$	$\vdots$	$\vdots$	$\vdots$	$\vdots$	$\vdots$	$\vdots$	$\vdots$	
64	320	1328	4864	16428	52356	159645	470010	...
32	144	560	1944	6266	19149	56190	159645	...
16	64	232	760	2329	6802	19149	52356	...
8	28	94	289	838	2329	6266	16428	...
4	12	37	106	289	760	1944	4864	...
2	5	14	37	94	232	560	1328	...
1	2	5	12	28	64	144	320	...
1	1	2	4	8	16	32	64	...

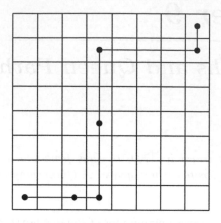

FIGURE 9.1: A Rook path.

It's elementary to obtain by inclusion–exclusion a recurrence formula for $a(m, n)$ that requires only a fixed number of previous terms, namely,

$$a(0,0) = 1, \ a(0,1) = 1, \ a(1,0) = 1, \ a(1,1) = 2;$$

$$a(m, n) = 2a(m, n-1) + 2a(m-1, n) - 3a(m-1, n-1), \ m \geq 2 \text{ or } n \geq 2.$$

We are assuming that $a(m, n) = 0$ for m or n negative.

We indicate this recurrence relation with an array of coefficients.

$$\begin{array}{cc} -2 & 1 \\ 3 & -2 \end{array}$$

The recurrence formula yields a rational generating function for the doubly-infinite sequence $\{a(m, n)\}$, namely,

$$\sum_{m \geq 0, \, n \geq 0} a(m, n) s^m t^n = \frac{1 - s - t + st}{1 - 2s - 2t + 3st}.$$

The form of the denominator of this function comes from the recurrence relation. The numerator is obtained by multiplying the denominator by the polynomial that represents the initial values, $1 + s + t + 2st$, and keeping only those monomials with exponents of s and t both less than 2.

We can pose the same kind of problem for a Rook moving in a 3-dimensional space. In how many ways can a Rook move from $(0, 0, 0)$ to (n, n, n), where each step is a positive integer multiple of $(1, 0, 0)$, $(0, 0, 1)$, or $(0, 0, 1)$?

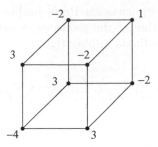

FIGURE 9.2: Coefficients of a recurrence relation for 3-D Rook paths.

The coefficients for a recurrence relation are indicated in Figure 9.2.

In any dimension, a Rook path is at each step a positive integer multiple of a vector that consists of a single 1 and the other coordinates 0. The recurrence relation for a d-dimensional Rook depends on decrementing each variable by 0 or 1, and each coefficient is equal to $(n+1)(-1)^n$, where n is the number of variables decremented.

For 3-dimensional Rook paths, the generating function is

$$\frac{(1-s)(1-t)(1-u)}{1 - 2(s+t+u) + 3(st+su+tu) - 4stu}.$$

We notice a pattern in both the numerator and the denominator. The numerator is the product of terms of the form $(1-x)$, where x is an indeterminate. The denominator is an alternating sum of elementary symmetric polynomials in three indeterminates. Let's describe the general situation. We are interested in counting lattice paths in d dimensions from the origin to a given point $p = (p_1, \ldots, p_d)$, such that each step is a positive integer multiple of a basic step of the form $u_i = (u_{i1}, \ldots, u_{id})$, where $1 \leq i \leq k$. As a convenient notation, let $x^\alpha = x_1^{\alpha_1} \ldots x_d^{\alpha_d}$, where $\alpha = (\alpha_1, \ldots, \alpha_d)$. For $0 \leq j \leq n$, the jth *elementary symmetric polynomial* in the indeterminates $x_1, \ldots, x_n$ is the sum of all products of j of the x_i.

Theorem 9.1. For $d \geq 1$ and $1 \leq i \leq k$, let $u_i = (u_{i1}, \ldots, u_{id})$ be a nonzero d-tuple of nonnegative integers. Let σ_j be the jth elementary symmetric polynomial in the indeterminates x^{u_i}. Then the number of lattice paths in d dimensions that start at $(0, \ldots, 0)$, stop at $p = (p_1, \ldots, p_d)$, where the p_i are nonnegative integers, and each step is a positive integer multiple of one of the u_i, is the coefficient of x^p in the rational generating function

$$\frac{\prod_{i=1}^k (1 - x^{u_i})}{\sum_{j=0}^k (-1)^j (j+1)\sigma_j}.$$

Proof. In the case of d-dimensional Rook paths, the recurrence relation which says that each value of the sequence is determined by adding its predecessors in each coordinate implies that

$$\sum_{(n_1,\ldots,n_d)} a(n_1,\ldots,n_d)x_1^{n_1}\ldots x_d^{n_d}\left(1-\sum_{i=1}^{d}\sum_{h=1}^{\infty}x_i^h\right) = a(0,\ldots,0) = 1.$$

Hence, using geometric series, we obtain

$$\sum_{(n_1,\ldots,n_d)} a(n_1,\ldots,n_d)x_1^{n_1}\ldots x_d^{n_d} = \frac{1}{1-\sum_{i=1}^{d}(x_i/(1-x_i))}$$

$$= \frac{\prod_{i=1}^{d}(1-x_i)}{\prod_{i=1}^{d}(1-x_i)-\sum_{j=1}^{d}x_j\prod_{i\neq j}(1-x_i)}.$$

The numerator is of the correct form. The summands in the denominator are products of distinct x_i. Let's find the coefficient of the product of some specific j of the x_i. The term $\prod_{i=1}^{d}(1-x_i)$ yields a coefficient of $(-1)^j$ and the term $-\sum_{j=1}^{d}x_j\prod_{i\neq j}(1-x_i)$ contributes a coefficient of $j(-1)^j$. Hence, the sum of the coefficients is $(j+1)(-1)^j$, as required.

The general result follows upon letting $x_i = x^{u_i}$, for $1 \leq i \leq k$. The fact that d is no longer necessarily the dimension of the space is immaterial. ∎

Let's apply this theorem to the problem of counting Queen paths. A chess Queen can move any number of squares horizontally, vertically, or diagonally in one step. In how many ways can a Queen move from the lower-left corner to the upper-right corner of an 8×8 chessboard, assuming that the Queen moves up, right, or diagonally up-right at each step? Such a Queen path is shown in Figure 9.3.

As with Rook paths, we can fill out a table of the number of Queen paths to each square. Let $b(m,n)$ be the number of paths from $(0,0)$ to (m,n), such that at each step the path goes up, right, or up-right. Notice that we calculate each entry by adding all the entries to the left of, below, and diagonally left-below that entry. For example, $b(3,3) = 4 + 17 + 60 + 4 + 17 + 60 + 1 + 3 + 22 = 188$. The reason for this rule is that the Queen has to arrive from one of the aforementioned squares. We see that the number of paths to the upper-right corner is $b(7,7) = 1499858$.

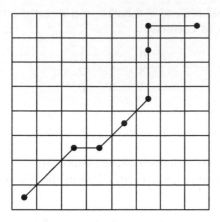

FIGURE 9.3: A Queen path.

64	464	2392	10305	39625	140658	470233	1499858	...
32	208	990	3985	14430	48519	154352	470233	...
16	92	401	1498	5079	16098	48519	140658	...
8	40	158	543	1712	5079	14430	39625	...
4	17	60	188	543	1498	3985	10305	...
2	7	22	60	158	401	990	2392	...
1	3	7	17	40	92	208	464	...
1	1	2	4	8	16	32	64	...

By Theorem 9.1, the generating function for the doubly-infinite sequence for Queen paths (with generators s, t, and st) is

$$\frac{(1-s)(1-t)(1-st)}{1 - 2(s + t + st) + 3(s \cdot t + s \cdot st + t \cdot st) - 4(s \cdot t \cdot st)}.$$

From the denominator, we obtain a recurrence formula:

$b(0,0) = 1,\ b(0,1) = 1,\ b(0,2) = 2,$

$b(1,0) = 1,\ b(1,1) = 3,\ b(1,2) = 7,$

$b(2,0) = 2,\ b(2,1) = 7,\ b(2,2) = 22;$

$b(m,n) = 2b(m-1,n) + 2b(m,n-1) - b(m-1,n-1) - 3b(m-2,n-1)$

$\qquad - 3b(m-1,n-2) + 4b(m-2,n-2), \quad m \geq 2 \text{ or } n \geq 2.$

We are assuming that $b(m, n) = 0$ for m or n negative.

Here is the array of coefficients.

$$\begin{array}{rrr} 0 & -2 & 1 \\ 3 & 1 & -2 \\ -4 & 3 & 0 \end{array}$$

An inclusion–exclusion proof for the recurrence relation is possible but tricky. The generating function approach is much easier.

Let $a_n = a(n, n)$, the nth diagonal element of the sequence for 2-D Rook paths. Then the sequence $\{a_n\}$ is A051708 in the Encyclopedia of Integer Sequences (EIS):

$$1, \ 2, \ 14, \ 106, \ 838, \ 6802, \ 56190, \ 470010, \ \ldots.$$

(In the database, the first term is $a(1)$, whereas ours is a_0.)

In 2003 Curtis Coker found the generating function and recurrence relation for the diagonal sequence for 2-D Rook paths. We will derive these here. We'll show that the generating function for the sequence is

$$f(x) = \frac{1}{2}\left(1 + \frac{(1 - x)}{\sqrt{(1 - x)(1 - 9x)}}\right).$$

In order to accomplish this, we make the change of variables $t = x/s$ (so that $st = x$). Now we allow arbitrary integer exponents of s, while the exponents of x are nonnegative integers. Thus, for example, we represent $s^3 t^5$ as $s^{-2} x^5$. The generating function becomes

$$\frac{1}{2}\left(1 + \frac{(1 - x)s}{-2s^2 + (3x + 1)s - 2x}\right).$$

We focus on the function

$$\frac{s}{-2s^2 + (3x + 1)s - 2x} = \frac{s}{-2(s - \alpha)(s - \beta)},$$

where

$$\alpha = \frac{3x + 1 - \sqrt{(1 - x)(1 - 9x)}}{4}, \quad \beta = \frac{3x + 1 + \sqrt{(1 - x)(1 - 9x)}}{4}.$$

The diagonal generating function is the coefficient of s^0.

The partial fractions expansion of our formula is

$$\frac{1}{2(\beta - \alpha)}\left[\frac{\alpha}{s - \alpha} - \frac{\beta}{s - \beta}\right] = \frac{1}{2(\beta - \alpha)}\left[\frac{\alpha/s}{1 - (\alpha/s)} + \frac{1}{1 - (s/\beta)}\right].$$

Expanding the function in a Laurent series in the annulus $|\alpha| < |s| < |\beta|$ in powers of (α/s) and (s/β), where $-1/9 < x < 1/9$, yields

$$\frac{1}{2(\beta - \alpha)} \left[\sum_{n=1}^{\infty} \left(\frac{\alpha}{s}\right)^n + \sum_{n=0}^{\infty} \left(\frac{s}{\beta}\right)^n \right].$$

The coefficient of s^0 is

$$\frac{1}{2(\beta - \alpha)} = \frac{1}{\sqrt{(1 - x)(1 - 9x)}}.$$

This establishes the formula for the generating function $f(x)$.

In finding a recurrence relation for the numbers a_n, it's easier to work with the function

$$g(x) = 2f(x) - 1 = \frac{\sqrt{1 - x}}{\sqrt{1 - 9x}},$$

rather than the generating function f. Note that f and g yield sequences satisfying the same recurrence relation but with different initial values. By logarithmic differentiation, we obtain

$$\log g(x) = \frac{1}{2} \log(1 - x) - \frac{1}{2} \log(1 - 9x),$$

and hence

$$\frac{g'(x)}{g(x)} = \frac{-\frac{1}{2}}{1 - x} + \frac{\frac{9}{2}}{1 - 9x},$$

or

$$g'(x)(1 - x)(1 - 9x) = 4g(x).$$

Sequences such as ours, whose generating functions consist of polynomials and a finite number of derivatives, are called *D-finite*.

We can read off the recurrence formula for the a_n directly:

$$a_0 = 1; \; a_1 = 2;$$

$$a_n = ((10n - 6)a_{n-1} - (9n - 18)a_{n-2})/n, \quad n \geq 2.$$

We do not know a counting proof of this formula.

Rook paths have a tie-in with a game called Nim. Nim is played with a number of piles of stones. Players alternately remove any number of stones from a single pile. The game ends when some player takes the last stone(s). Rook paths in d dimensions that go from $(0, 0, \ldots, 0)$ to $(a_1, a_2, \ldots, a_d)$ are equivalent to Nim games that start with d piles of stones of sizes $a_1, a_2, \ldots, a_d$. Therefore, by our analysis, the number of Nim games that start with two equal piles of n stones satisfies a linear recurrence relation of order 2, with polynomial coefficients of degree 1.

In 2009 Khang Tran, Suren Fernando, and the author found that the number of d-dimensional Rook paths from $(0, \ldots, 0)$ to $(n, \ldots, n)$ is asymptotic to

$$(d+1)^{dn-1} d^{(d+2)/2} (2\pi n(d+2))^{(1-d)/2}.$$

This is also the asymptotic number of Nim games that start with d piles of n stones each.

We now return to Queen paths. The diagonal sequence for the 2-D Queen, $\{b_n = b(n,n)\}$, is the EIS sequence A132595:

1, 3, 22, 188, 1712, 16098, 154352, 1499858, 14717692, 145509218, ...

Let $x = st$. Then the generating function becomes

$$\frac{(x-1)(s^2 - (-x-1)s + x)}{(3x-2)s^2 + (-4x^2 + x + 1)s + (3x^2 - 2x)}.$$

We focus on the function

$$\frac{s^2 + (-x-1)s + x}{(3x-2)s^2 + (-4x^2 + x + 1)s + (3x^2 - 2x)}$$

$$= \frac{1}{3x-2} + \frac{(x-1)^2}{(3x-2)^2} \cdot \left[\frac{1}{\alpha - \beta} \left[\frac{\alpha}{s - \alpha} - \frac{\beta}{s - \beta} \right] \right],$$

where

$$\alpha, \beta = \frac{1}{2} \left[\frac{4x^2 - x - 1}{3x - 2} \mp \sqrt{\Delta} \right]$$

$$\Delta = \frac{(x-1)^2 (1 - 12x + 16x^2)}{(3x-2)^2}.$$

We rewrite this function as

$$\frac{1}{3x-2} + \frac{(x-1)^2}{(3x-2)^2} \left[\frac{1}{\alpha - \beta} \left[\frac{\alpha/s}{1 - (\alpha/s)} + \frac{1}{1 - (s/\beta)} \right] \right].$$

Expanding in $|\alpha| < |s| < |\beta|$, we obtain

$$\frac{1}{3x-2} + \frac{(x-1)^2}{(3x-2)^2} \left[\frac{1}{\alpha - \beta} \left[\sum_{n=1}^{\infty} (\alpha/s)^n + \sum_{n=0}^{\infty} (s/\beta)^n \right] \right].$$

Therefore

$$f(x) = (x-1)\left[\frac{1}{3x-2} + \frac{(x-1)^2}{(3x-2)^2} \cdot \frac{-1}{\sqrt{\Delta}}\right]$$

$$= \frac{(x-1)}{(3x-2)}\left[1 + \frac{1-x}{\sqrt{1-12x+16x^2}}\right].$$

Solving for $1/\sqrt{1-12x+16x^2}$ and taking a derivative yields

$$f(x)(46x^2-47x+11)+f'(x)(48x^4-116x^3+95x^2-29x+2) = 10x^2-15x+5,$$

and we can read off the recurrence formula:

$$b_0 = 1; \ b_1 = 3; \ b_2 = 22; \ b_3 = 188;$$

$$b_n = ((29n-18)b_{n-1} + (-95n+143)b_{n-2}$$

$$+ (116n-302)b_{n-3} + (-48n+192)b_{n-4})/(2n), \quad n \ge 4.$$

In any dimension, a Queen path is at each step a positive integer multiple of a vector consisting of only 0s and 1s (and not all 0s).

Wythoff's Nim, named after Willem Abraham Wythoff (1865–1939), is a game that starts with a number of piles of stones. Players alternately remove the same number of stones from any number of piles. The game ends when a player takes the last stone(s). Queen paths in d dimensions that go from $(0, 0, \ldots, 0)$ to $(a_1, a_2, \ldots, a_d)$ are equivalent to Wythoff's Nim games that start with d piles of stones of sizes $a_1, a_2, \ldots, a_d$. Therefore, the number of Wythoff's Nim games that start with two equal piles of n stones satisfies a linear recurrence relation of order 4, with polynomial coefficients of degree 1.

Exercises

1. A lone King is on a chessboard. How many ways may the King travel from the lower-left corner of the board to the upper-right corner, moving one square right, up, or diagonally up-right at every step?

2. Show that the number of Rook paths from $(0,0)$ to $(n,0)$ is 2^{n-1}. For what well-known counting problem is this the formula?

◇3. How many 3-D Rook paths are there from $(0,0,0)$ to $(7,7,7)$?

⋄4. Write the generating function and recurrence relation coefficients for a depth four linear recurrence relation for the 3-D Queen. How many Queen paths are there from $(0,0,0)$ to $(7,7,7)$?

5. Suppose that a piece moves in the plane like a Queen but only diagonally and horizontally (not vertically). Write the generating function and recurrence relation for the number of paths that this piece can take going from $(0,0)$ to (m,n), always moving right or up-right at each step.

⋄6. You have two stamp rolls, one with 1-cent stamps and the other with 2-cent stamps. Let $a(n)$ be the number of ways to make postage of n cents by taking strips of stamps from the two rolls. The order of the strips and the number of stamps per strip matter. For example, $a(4) = 15$ since

$$4 = (1) + (1) + (1) + (1) = (1+1) + (1) + (1)$$
$$= (1) + (1+1) + (1) = (1) + (1) + (1+1)$$
$$= (1+1) + (1+1) = (1+1+1) + (1)$$
$$= (1) + (1+1+1) = (1+1+1+1)$$
$$= (2) + (1) + (1) = (1) + (2) + (1) = (1) + (1) + (2)$$
$$= (2) + (1+1) = (1+1) + (2) = (2) + (2) = (2+2).$$

Find $a(100)$, the number of ways to make postage of \$1. Give an approximation of $a(10^5)$, the number of ways to make postage of \$1000.

7. Professor Bumble announces that there is only one number n for which $a(n)$ is a prime number. (See the previous problem.) What is n and can you prove that it is the only such number?

⋆8. Find a direct formula (not a recurrence relation) for $a(m,n)$, the number of Rook paths from $(0,0)$ to (m,n).

⋆9. Suppose that a ChildRook moves like a chess Rook but only at most two squares horizontally or vertically at each step. Let $a(m,n)$ be the number of ways that a ChildRook can move from $(0,0)$ to (m,n). Assume that the ChildRook always moves right or up at each step.

(a) Find a finite-order recurrence relation for $\{a(m,n)\}$.

(b) Find a rational generating function for $\{a(m,n)\}$.

(c) Find a recurrence formula for the diagonal sequence $\{a(n,n)\}$.

*10. Suppose that a RookPlus moves like a chess Rook with the additional option of moving one square diagonally at each step. Let $b(m, n)$ be the number of ways that a RookPlus can move from $(0, 0)$ to (m, n). Assume that the RookPlus always moves right, up, or up-right at each step.

  (a) Find a finite-order recurrence relation for $\{b(m, n)\}$.

  (b) Find a rational generating function for $\{b(m, n)\}$.

  (c) Find a recurrence formula for the diagonal sequence $\{b(n, n)\}$.

*11. Suppose that a HalfRook moves like a chess Rook, any number of squares horizontally, but only one square up at each step. Let $c(m, n)$ be the number of ways that a HalfRook can move from $(0, 0)$ to (m, n). Assume that the HalfRook always moves right or up at each step.

  (a) Find a finite-order recurrence relation for $\{c(m, n)\}$.

  (b) Find a rational generating function for $\{c(m, n)\}$.

  (c) Find a recurrence formula for the diagonal sequence $\{c(n, n)\}$.

*12. Suppose that a BishopPlus moves like a chess Bishop, any number of squares diagonally, or one square horizontally or vertically at each step. Let $d(m, n)$ be the number of ways that a BishopPlus can move from $(0, 0)$ to (m, n). Assume that the BishopPlus always moves right, up, or up-right at each step.

  (a) Find a finite-order recurrence relation for $\{d(m, n)\}$.

  (b) Find a rational generating function for $\{d(m, n)\}$.

  (c) Find a recurrence formula for the diagonal sequence $\{d(n, n)\}$.

 13. Construct a generating function for the number of 2-D Rook paths that also keeps track of the number of steps. Let $a(m, n; k)$ be the number of Rook paths from $(0, 0)$ to (m, n) that take k steps. Find a recurrence relation for $a(m, n; k)$.

 14. Generate a table for the number of paths of a generalized Knight from $(0, 0)$ to a point (m, n). The generalized Knight may move in any positive multiple of the basic steps $(1, 2)$ or $(2, 1)$. What pattern do you find? What is the explanation?

◇15. How many Nim games can be played starting with three piles of stones of sizes 10, 10, and 20? How many games of Wythoff's Nim?

◇16. Let a_n be the number of 3-D Rook paths from $(0, 0, 0)$ to (n, n, n), i.e., lattice paths in which each step is a positive integer multiple of $(1, 0, 0)$, $(0, 1, 0)$, or $(0, 0, 1)$. Find a linear recurrence formula with polynomial coefficients for $\{a_n\}$.

Part IV

Discrete Probability

Chapter 10

Probability Spaces and Distributions

In a game of bridge, all 52 playing cards are dealt randomly to four players, 13 cards per player. The probability that at least one player has all cards of the same suit (a perfect hand) is

$$18772910672458601/7450658022984554561000520000 \doteq 2.5 \times 10^{-11}.$$

Suppose that an urn contains n white balls and n black balls. A ball is selected at random from the urn and removed. This process is repeated until the urn contains only balls of one color. If n is large, then the expected number of balls remaining in the urn is approximately 2.

A *sample space* is a set of (simple) events or outcomes of an experiment.

Example 10.1. Consider one flip of a fair coin. What is the sample space?

Solution: The possible outcomes are heads and tails. Thus, we can represent the sample space as $\{H, T\}$. ■

A *probability space* is a sample space with a probability $\Pr(E)$ assigned to each event E. The probability of each simple event is nonnegative and the sum of the probabilities is 1.

Example 10.2. In our coin example, what is the probability space?

Solution: The events H and T each occur with probability $1/2$, so we write

$$\Pr(H) = \frac{1}{2} \quad \text{and} \quad \Pr(T) = \frac{1}{2}.$$

We can also consider combinations of the simple events. For instance, one of H and T must occur, so

$$\Pr(\{H, T\}) = 1.$$

The complementary probability is 0, that is,

$$\Pr(\emptyset) = 0.$$

■

67

Example 10.3. A single card is dealt from a deck of 52 cards. What is the probability space?

Solution: The sample space consists of all 52 cards, each with probability 1/52 of being chosen. ∎

In discrete mathematics, we are mainly concerned with sample spaces with only a finite number of elements. However, sometimes the sample space may be infinite if its elements are discrete, i.e., they are countable. The following is an example of a sample space with a countably infinite number of elements.

Example 10.4. A fair coin is flipped until heads occurs. What is the probability space?

Solution: The sample space consists of all strings of T's (including the empty string) followed by a single H. Thus, the sample space is the countably infinite set

$$\{H, TH, TTH, TTTH, TTTTH, TTTTTH, TTTTTTH, \ldots\}.$$

For $n \geq 0$, the probability of a string of n tails occurring followed by heads is $(1/2)^n(1/2) = (1/2)^{n+1}$. Hence, the sum of all the probabilities is

$$\sum_{n=0}^{\infty} \left(\frac{1}{2}\right)^{n+1} = 1.$$

∎

A *random variable* defined on a probability space is a function from the set of possible events to the set of real numbers.

For instance, suppose that a fair coin is tossed and we record 10 points if the coin lands heads and -3 points if it lands tails. Then the random variable X given by this situation is

$$X = \begin{cases} 10 & \text{with probability } 1/2 \\ -3 & \text{with probability } 1/2. \end{cases}$$

Let X be a random variable that takes values x_i with probabilities p_i, for $1 \leq i \leq n$. The *mean*, or *expected value*, of X is

$$\mu(X) = E(X) = \sum_{i=1}^{n} p_i x_i.$$

The *variance* of X is

$$V(X) = E((X - E(X))^2) = E(X^2) - E(X)^2 = \sum_{i=1}^{n} p_i x_i^2 - \mu^2.$$

The *standard deviation* of X is

$$\sigma(X) = \sqrt{V(X)}.$$

A *Bernoulli* random variable, named after Jacob Bernoulli (1654–1705), is a random variable X such that

$$X = \begin{cases} 1 & \text{with probability } p \\ 0 & \text{with probability } q, \end{cases}$$

where $0 \leq p \leq 1$ and $p + q = 1$. We say that 1 represents "success" and 0 represents "failure." We denote by $B(p)$ the Bernoulli variable with success probability p.

It is easy to determine the expected value (i.e., the mean) and variance of a Bernoulli variable:

$$E(X) = 0 \cdot q + 1 \cdot p = p,$$

and

$$V(X) = (q \cdot 0^2 + p \cdot 1^2) - p^2 = pq.$$

We say that two variables X and Y are *independent* if knowledge of the value of one of them yields no information about the distribution of the other.

Proposition 10.5. If $X_1, X_2, \ldots, X_n$ are random variables with sum X, then

$$E(X) = E(X_1) + E(X_2) + \cdots + E(X_n).$$

If the X_i are independent, then

$$V(X) = V(X_1) + V(X_2) + \cdots + V(X_n).$$

Suppose that $X_1, \ldots, X_n$ are independent Bernoulli random variables with success probability p. Let $X = X_1 + \cdots + X_n$. We call X a *binomial random variable*, and denote it by $B(n, p)$.

We have

$$E(X) = E(X_1) + \cdots + E(X_n)$$

$$= p + \cdots + p$$

$$= np$$

and

$$V(X) = V(X_1) + \cdots + V(X_n)$$

$$= pq + \cdots + pq$$

$$= npq.$$

The possible values of X are $0, 1, \ldots, n$. We have $X = k$ only when k of the X_i are equal to 1 (and the others are equal to 0), and this can happen in $\binom{n}{k}$ ways. Therefore

$$\Pr(X = k) = \binom{n}{k} p^k q^{n-k}, \quad 0 \le k \le n.$$

In Example 3.7, we gave three proofs of the identity

$$\sum_{k=1}^{n} k \binom{n}{k} = n2^{n-1}.$$

Let's also give a proof using probability. Upon division by 2^n, our identity becomes

$$\sum_{k=1}^{n} k \binom{n}{k} \left(\frac{1}{2}\right)^n = \frac{n}{2}.$$

Here is a probabilistic interpretation. Let S be a set of n elements. For each element of S, flip a fair coin and if the coin comes up heads put the element in a subset T. What is the expected size of T? Both sides of the identity give the answer.

The inclusion–exclusion principle is a generalization of the Venn diagram rule.

Proposition 10.6 (Venn Diagram Rule). If A and B are finite sets, then

$$|A \cup B| = |A| + |B| - |A \cap B|.$$

Proof. Figure 10.1 shows two sets, A and B, and their union and intersection. The sum $|A| + |B|$ counts all the elements of $A \cup B$, but the elements of $A \cap B$ are counted twice and therefore must be removed as on the right side of the relation. ∎

$$A \cup B$$

FIGURE 10.1: Venn diagram for two sets.

Theorem 10.7 (Inclusion–Exclusion Principle). If $A_1, \ldots, A_n$ are finite sets, then

$$|A_1 \cup \cdots \cup A_n| = \sum_{i=1}^{n} (-1)^{i+1} \sum_{1 \leq k_1 < \cdots < k_i \leq n} |A_{k_1} \cap \cdots \cap A_{k_i}|.$$

Proof. Let S be a finite set containing all the A_i. For $s \in S$, suppose that s is contained in exactly m of the A_i. If $m = 0$, then the contribution of s to the right side of our relation is 0. If $m \geq 1$, then the contribution is

$$\sum_{j=1}^{n} (-1)^{j+1} \binom{m}{j} \;=\; \sum_{j=1}^{m} (-1)^{j+1} \binom{m}{j} \qquad \text{(because } m \leq n)$$

$$= \; 1 \qquad \text{(by Proposition 3.1).}$$

Hence, each $s \in S$ not in the union of the A_i contributes 0 to both sides of the relation, while each $s \in S$ in the union contributes 1. Therefore, each element of S contributes an equal amount to both sides of the relation. This confirms the formula. ∎

Example 10.8. A *derangement* of a set is a permutation of the set with no fixed points. Let d_n be the number of derangements of n elements. Find a formula for d_n.

Solution: For $1 \leq j \leq n$, let A_j be the set of permutations of $\{1, 2, 3, \ldots, n\}$ such that j is a fixed point. Then the intersection of any i of the A_j, for $1 \leq i \leq n$, has $(n - i)!$ elements, because the $n - i$ not necessarily fixed elements may be permuted arbitrarily. Since $\binom{n}{i}$ of the A_j make up the intersection, by the principle of inclusion and exclusion, we have

$$|A_1 \cup \cdots \cup A_n| \;=\; \sum_{j=1}^{n} (-1)^{j+1} \binom{n}{j} (n-j)!.$$

A permutation is a derangement if it is not a member of one of the A_j, so we conclude that

$$d_n = \sum_{j=0}^{n} (-1)^j \frac{n!}{j!}.$$

∎

Theorem 10.9 (Inclusion–Exclusion Principle, Probability Version). Let $E_1, \ldots, E_n$ be events in a finite probability space. Then

$$\Pr(E_1 \cup \cdots \cup E_n) \;=\; \sum_{i=1}^{n} (-1)^{i+1} \sum_{1 \leq k_1 < \cdots < k_i \leq n} \Pr(E_{k_1} \cap \cdots \cap E_{k_i}).$$

Example 10.10. A deck of 52 playing cards is dealt randomly to four players, 13 cards per player. What is the probability that at least one player has all cards of the same suit?

Solution: For $1 \leq i \leq 4$, let E_i be the event that player i has all cards of the same suit. Using the principle of inclusion and exclusion, we find that the desired probability, $\Pr(\bigcup E_i)$, is

$$\binom{4}{1} \frac{4}{\binom{52}{13}} - \binom{4}{2} \frac{\binom{4}{2}2!}{\binom{52}{13}\binom{39}{13}} + \binom{4}{3} \frac{\binom{4}{3}3!}{\binom{52}{13}\binom{39}{13}\binom{26}{13}} - \binom{4}{4} \frac{\binom{4}{4}4!}{\binom{52}{13}\binom{39}{13}\binom{26}{13}\binom{13}{13}}$$

$$= 18772910672458601/745065802298455456100520000$$

$$\doteq 2.5 \times 10^{-11}.$$

To understand the above calculation, consider for example the second term,

$$\binom{4}{2} \frac{\binom{4}{2}2!}{\binom{52}{13}\binom{39}{13}}.$$

This is the probability that at least two players have all cards of the same suit. The leading expression $\binom{4}{2}$ counts the choices of the two lucky players. The term $\binom{4}{2}$ in the numerator counts the choices of the two solid suits. This expression is multiplied by 2! to account for the ways that the two lucky players can have the two chosen suits. The denominator is the number of ways of dealing 13 cards each to the two lucky players. ∎

Example 10.11. What is the probability P_n that a random permutation of n elements is a derangement?

Solution: We found in Example 10.8 that

$$d_n = \sum_{j=0}^{n} (-1)^j \frac{n!}{j!}.$$

Therefore

$$P_n = \sum_{j=0}^{n} \frac{(-1)^j}{j!}.$$

Thus, $\{P_n\}$ is a zigzag sequence (consecutive values rise and fall) and

$$\lim_{n \to \infty} P_n = e^{-1} \doteq 0.37.$$

∎

Example 10.12. Show that the expected number of fixed points of a permutation of n elements is 1.

Solution: We illustrate the result in the case $n = 3$. Below are the permutations of $\{1, 2, 3\}$ and the number of fixed points of each.

permutation	number of fixed points
(1)(2)(3)	3
(1)(2 3)	1
(2)(1 3)	1
(3)(1 2)	1
(1 2 3)	0
(1 3 2)	0

The permutations are written in cycle form. For instance, the permutation $(2)(13)$ is the one that maps $2 \to 2$, $1 \to 3$, and $3 \to 1$. The total number of fixed points is 6 and the average is $6/6 = 1$.

Randomly choose a permutation of $\{1, 2, \ldots, n\}$. For $1 \le i \le n$, define $X_i = 1$ if i is fixed and 0 otherwise. Then the number of fixed points is $X = X_1 + X_2 + \cdots + X_n$. The expected value of each X_i is $(n-1)!/n! = 1/n$. Hence, the expected number of fixed points is

$$E(X) = E(X_1) + E(X_2) + \cdots + E(X_n)$$

$$= \frac{1}{n} + \frac{1}{n} + \cdots + \frac{1}{n}$$

$$= n \cdot \frac{1}{n}$$

$$= 1.$$

■

Example 10.13. An object travels along the integer points of the plane, starting at the point $(0, 0)$. At each step, the object moves one unit to the right or one unit up (with equal probability). The object stops when it reaches the line $x = n$ or the line $y = n$. Show that the expected length of the object's lattice path is

$$2n - n \binom{2n}{n} 2^{1-2n}.$$

Solution: Assume that the object hits the line $x = n$ at the point (n, k) or the line $y = n$ at the point (k, n), where $0 \leq k \leq n - 1$. Then the expected length of the path is given by

$$E = \sum_{k=0}^{n-1} (n+k)2 \cdot \frac{1}{2} \binom{n+k-1}{n-1} \left(\frac{1}{2}\right)^{n+k-1}$$

$$= \left(\frac{1}{2}\right)^n \sum_{k=0}^{n-1} (n+k) \binom{n+k-1}{n-1} \left(\frac{1}{2}\right)^{k-1}$$

$$= \left(\frac{1}{2}\right)^n n \sum_{k=0}^{n-1} \binom{n+k}{n} \left(\frac{1}{2}\right)^{k-1}$$

$$= \left(\frac{1}{2}\right)^n 2n \sum_{k=0}^{n-1} \binom{n+k}{n} \left(\frac{1}{2}\right)^{k}$$

$$= \left(\frac{1}{2}\right)^n 2n \left(\sum_{k=0}^{n} \binom{n+k}{n} \left(\frac{1}{2}\right)^{k} - \binom{2n}{n} \left(\frac{1}{2}\right)^{n} \right).$$

By the result of Example 3.4, this simplifies to

$$E = 2n - n \binom{2n}{n} 2^{1-2n}.$$

∎

The inclusion–exclusion principle is generalized by the Bonferroni inequalities of probability theory, named after Carlo Emilio Bonferroni (1892–1960).

Theorem 10.14 (Bonferroni Inequalities). Let $A_1, \ldots, A_n$ be subsets of a finite set S. If t is an odd positive integer, then

$$|A_1 \cup \cdots \cup A_n| \leq \sum_{i=1}^{t} (-1)^{i+1} \sum_{1 \leq k_1 < \cdots < k_i \leq n} |A_{k_1} \cap \cdots \cap A_{k_i}|.$$

If t is even, then the inequality is reversed.

Proof. Let $s \in S$ and assume that s is contained in exactly m of the A_i. If $m = 0$, then the contribution to both sides of the inequality is 0. For $m > 0$, the result follows from the identity $\sum_{i=0}^{t} \binom{m}{i}(-1)^i = \binom{m-1}{t}(-1)^t$. (See Exercise 7 of Chapter 3.) ∎

Example 10.15. Use the Bonferroni inequalities to give bounds on d_n, the number of derangements of $\{1, 2, 3, \ldots, n\}$.

Solution: If k is even, then

$$d_n \leq \sum_{j=0}^{k} (-1)^j \frac{n!}{j!}.$$

If k is odd, then the inequality is reversed. ∎

The Bonferroni inequalities have a natural probability interpretation when both sides of an inequality are divided by the size of the sample space. For instance, the probability that a random permutation of n elements is a derangement is bounded by

$$\sum_{j=0}^{k} \frac{(-1)^j}{j!}.$$

This is an upper bound if k is even and a lower bound if k is odd.

Let's consider some other useful probability distributions.

The distribution of the *negative binomial random variable* is defined to be the probability of obtaining, in a sequence of $s + f$ Bernoulli trials, s successes and f failures, with a success on the last trial. This probability is

$$\binom{s+f-1}{f} p^s q^f, \quad s \geq 1, f \geq 0.$$

Notice that the coefficient above is the coefficient of x^f in the series expansion of $(1+x)^{-s}$ (see p. 14).

Example 10.16. A person shooting basketball free throws has a 0.7 chance of success. What is the probability that in 10 attempts the shooter will have seven successes and three failures, with a success on the last attempt?

Solution: The probability is given by a negative random variable with $p = 0.7$, $q = 0.3$, $s = 7$, and $f = 3$. Thus, the probability is

$$\binom{9}{3} (0.7)^7 (0.3)^3 = 0.18678.$$

∎

In the special case $s = 1$, we have the *geometric random variable*, with distribution given by

$$pq^f, \quad f \geq 0.$$

Example 10.17. Another free throw shooter has a 0.1 chance of success. In 10 attempts, what is the probability that the shooter misses the first nine and makes the last one?

Solution: The probability is given by a geometric random variable with $p = 0.1$, $q = 0.9$, and $f = 9$. Thus, the probability is

$$(0.1)(0.9)^9 = 0.0909533.$$

∎

The *hypergeometric random variable* has a distribution given by the probability of obtaining a certain sample from a bin of marbles. Suppose that a bin contains b blue marbles and g green marbles. A sample of size m is made without replacement. The probability that exactly k blue marbles are selected is

$$\frac{\binom{b}{k}\binom{g}{m-k}}{\binom{b+g}{m}}, \quad 0 \le m \le b+g, \ 0 \le k \le b.$$

The fact that the sum of these probabilities is 1 can be seen from Vandermonde's identity.

Example 10.18. An urn contains 10 green marbles and 10 blue marbles. A random selection of 10 balls is made from the urn without replacement. What is the probability that the selection consists of five green marbles and five blue marbles?

Solution: The probability is

$$\frac{\binom{10}{5}\binom{10}{5}}{\binom{20}{10}} = \frac{15876}{46189} \doteq 0.343718.$$

∎

Example 10.19. Suppose that an urn contains n white balls and n black balls. A ball is selected at random from the urn and removed. This process is repeated until the urn contains only balls of one color. What is the expected number of balls remaining in the urn?

Solution: What is the probability that a given white ball is left in the urn? Imagine that we continue the selection and withdrawal process until the urn is empty. Then the given white ball is left in the urn (in the original set-up) if and only if it is selected after all the black balls in the extended process. Relative to the black balls, and ignoring all the other white balls, there are $n + 1$ places at which the white ball can be selected: before the first black ball, between the first and second black balls, ..., after the last black ball. Since these places are equally likely, the probability that the given white ball is left in the urn is $1/(n + 1)$. As each of the $2n$ balls (black or white) has this probability of remaining, the expected number of balls remaining is $2n/(n+1)$.

As n tends to infinity, the expected number of balls left in the urn approaches 2. ■

We end this chapter with a brief description of some central results in probability theory, most importantly, the law of large numbers. The variables in our discussion may be continuous or discrete.

The basic result is due to Andrey Markov (1856–1922).

Lemma 10.20 (Markov's Inequality). Let X be a random variable with mean μ, and λ any positive real number. Then

$$\Pr(X \geq \lambda) \leq \mu/\lambda.$$

Proof. From the definition of μ, we obtain

$$\lambda \Pr(X \geq \lambda) \leq \mu.$$

The result follows immediately. ■

The next result is attributed to Pafnuty Lvovich Chebyshev (1821–1894).

Theorem 10.21 (Chebyshev's Inequality). Let X be a random variable with mean μ and variance σ^2. Then, for any real number $k > 0$, we have

$$\Pr(|X - \mu| \geq k\sigma) \leq \frac{1}{k^2}.$$

Proof. By Markov's inequality,

$$\sigma^2 = V(X) = E((X - \mu)^2) \geq k^2\sigma^2 \Pr(|X - \mu| \geq k\sigma).$$

■

Theorem 10.22 (Law of Large Numbers for Repeated Trials of a Bernoulli Random Variable). Let X be a Bernoulli variable. Then, for any positive number ϵ, we have

$$\Pr\left(|\overline{X} - E(\overline{X})| \geq \epsilon\right) \longrightarrow 0 \quad (\text{as } n \to \infty),$$

where $\overline{X} = (X_1 + \cdots + X_n)/n$, and the X_i are independent and identically distributed to X.

Proof. If X is a Bernoulli random variable of type $B(p)$, then $\overline{X}$ has standard deviation $\sigma = \sqrt{pq/n}$. By Chebyshev's inequality,

$$\Pr\left(|\overline{X} - E(\overline{X})| \geq k\sqrt{pq/n}\right) \leq \frac{1}{k^2}.$$

Choose k so that $k\sqrt{pq/n} \geq \epsilon$. Now

$$\frac{1}{k^2} \leq \frac{pq}{\epsilon^2 n} \longrightarrow 0 \quad (\text{as } n \to \infty).$$

∎

Exercises

1. A deck of 52 playing cards is dealt to two players so that each player receives half of the deck. What is the sample space in this situation and how many elements does it have? What is the probability of each simple event in the sample space?

2. A pair of dice is rolled. What is the sample space? What is the probability that the sum of the two dice is 7?

3. Three dice are rolled. What is the probability that the sum of the three dice is at least 17?

4. Let $X \sim B(n, 1/2)$, with n odd. Prove that $\Pr(X < n/2) = 1/2$.

5. An unfair coin has probability p of coming up heads ($p > 1/2$). We flip the coin repeatedly until it comes up heads. What is the expected number of flips?

6. Prove that the expected number of "runs" in a sequence of n independent Bernoulli variables $B(p)$ is $2p(1-p)(n-1) + 1$. (A run is a longest consecutive sequence of identical outcomes. For example, the sequence 001110010101 has eight runs.)

7. Prove that the expected number of different birth dates (out of 365 equally likely dates) among a group of n people is

$$365\left(1 - \left(\frac{364}{365}\right)^n\right).$$

†8. Find a recurrence formula for the derangement numbers d_n.

◇9. Use a computer and the result of the previous exercise to calculate d_{30}.

†10. Prove the following recurrence relation for the derangement numbers:

$$d_n = nd_{n-1} + (-1)^n, \quad n \geq 2.$$

†⋆11. Find a simple formula for the exponential generating function for $\{d_n\}$, that is, the function

$$\sum_{n=0}^{\infty} d_n \frac{x^n}{n!}.$$

⋆12. For $n \geq 1$, let d_n be the number of derangements of n elements. Then the expected number of fixed points in a permutation of n elements is

$$\frac{1}{n!} \sum_{k=1}^{n} k \binom{n}{k} d_{n-k}.$$

Show that this expression simplifies to 1.

13. Two decks of 52 playing cards are shuffled and then dealt face up from both decks one at a time. How many "matches" are expected? A match is the same card (rank and suit) dealt from both decks.

⋆14. All 52 playing cards are dealt to four players, 13 cards per player. What is the probability that exactly one player has all cards of the same suit?

†⋆15. Find a formula for Euler's function $\phi(n)$, which is the number of integers between 1 and n that have no common factor with n.

⋆16. Show that the formula in Example 10.13 tends to

$$2n \left(1 - \frac{1}{\sqrt{\pi n}} \right),$$

as $n \to \infty$.

⋆17. Show that a pair of dice cannot be weighted so as to give all sums 2, ..., 12 with equal probability.

⋆18. Can three dice be weighted so as to give all sums 3, ..., 18 with equal probability?

19. A juggler can do a certain routine with probability of success 0.9. Given 10 attempts, what is the probability that the juggler succeeds nine times with the last attempt a success?

20. An urn contains 10 green marbles and 20 blue marbles. A random selection of 10 balls is made from the urn without replacement. What is the probability that the selection consists of five green marbles and five blue marbles?

◇21. An urn contains 100 green marbles and 100 red marbles. A random selection of 100 marbles is made from the urn without replacement. Use a computer to find the probability that the number of green marbles selected is between 48 and 52.

22. An urn contains nine red balls, nine white balls, and nine blue balls. A random selection of nine balls is made from the urn without replacement. What is the probability that the selection consists of three balls of each color?

23. Suppose that an urn contains w white balls and b black balls. A ball is selected at random from the urn and removed. This process is repeated until the urn contains only balls of one color. What is the expected number of balls remaining in the urn?

24. An urn contains balls numbered $1, \ldots, n$ but otherwise identical. A ball is picked at random from the urn, its number noted, and then returned to the urn. This operation is performed three times. Prove that the probability that the sum of the three numbers obtained is divisible by 3 is at least $1/4$.

25. An urn contains five white balls and five black balls. A ball is withdrawn at random. If it is white it is returned to the urn. If it is black it is left out. The process is repeated until all the black balls have been removed from the urn. What is the expected number of balls withdrawn from the urn?

26. There are two urns. Urn A contains five white balls. Urn B contains four white balls and one black ball. An urn is selected at random and a ball in that urn is selected at random and removed. This procedure is repeated until one of the urns is empty. What is the probability that the black ball has not been selected?

Chapter 11

Markov Chains

A boy, a girl, and a dog are playing with a ball. The boy throws the ball to the girl 2/3 of the time and to the dog 1/3 of the time. The girl throws the ball to the boy 1/2 of the time and to the dog 1/2 of the time. The dog brings the ball to the girl all of the time. Then, on average, the boy will have the ball 3/13 of the time, the girl 6/13 of the time, and the dog 4/13 of the time.

Suppose that a boy, a girl, and a dog play with a ball according to the scenario described in the introduction. If play continues for a long time, and the probabilities that the boy, the girl, and the dog have the ball at any given time are b, g, and d, respectively, then these probabilities satisfy the equations

$$b = \frac{1}{2}g$$

$$g = \frac{2}{3}b + d$$

$$d = \frac{1}{3}b + \frac{1}{2}g$$

$$b + g + d = 1.$$

We can solve these equations using elementary algebra, finding that $b = 3/13$, $g = 6/13$, and $d = 4/13$. This is called the "steady-state solution" to the problem. We would like to show that the probabilities always tend to this steady-state solution.

We can turn the situation into a matrix equation. Let the probabilities that the boy, the girl, and the dog have the ball at time n be b_n, g_n, and d_n, respectively. The initial probabilities b_0, g_0, and d_0 are three nonnegative real numbers that sum to 1. These probabilities satisfy the matrix equations

$$\begin{bmatrix} b_{n+1} \\ g_{n+1} \\ d_{n+1} \end{bmatrix} = \begin{bmatrix} 0 & 1/2 & 0 \\ 2/3 & 0 & 1 \\ 1/3 & 1/2 & 0 \end{bmatrix} \begin{bmatrix} b_n \\ g_n \\ d_n \end{bmatrix}, \quad n \geq 0.$$

After the boy, girl, and dog have been playing with the ball for a long time, we expect that there are limiting values of the probabilities b_n, g_n, and d_n. Let

$$b = \lim_{n \to \infty} b_n, \quad g = \lim_{n \to \infty} g_n, \quad d = \lim_{n \to \infty} d_n.$$

We expect that b, g, and d satisfy the equation

$$\begin{bmatrix} b \\ g \\ d \end{bmatrix} = \begin{bmatrix} 0 & 1/2 & 0 \\ 2/3 & 0 & 1 \\ 1/3 & 1/2 & 0 \end{bmatrix} \begin{bmatrix} b \\ g \\ d \end{bmatrix}.$$

As we will see, these limits don't depend on b_0, g_0, and d_0, and they can be found using algebra as above.

A system such as the one described is called a Markov chain. The concept of a Markov chain is due to Andrey Markov (1856–1922), who introduced it in 1906. The matrix

$$P = \begin{bmatrix} 0 & 1/2 & 0 \\ 2/3 & 0 & 1 \\ 1/3 & 1/2 & 0 \end{bmatrix}$$

is called the transition matrix for the Markov chain.

A *Markov chain* is a sequence X_0, X_1, ... of random variables that indicate *states* 1, 2, 3, ..., m such that there is a probability p_{ij} that $X_{n+1} = i$ given that $X_n = j$. The p_{ij}, called *transition probabilities*, are nonnegative numbers whose sum, for each fixed i, is 1. We can write the transition probabilities in matrix form called a *transition matrix*:

$$P = \begin{bmatrix} p_{11} & \cdots & p_{1m} \\ p_{21} & \cdots & p_{2m} \\ \vdots & \ddots & \vdots \\ p_{m1} & \cdots & p_{mm} \end{bmatrix}.$$

A state vector X is changed to a new state vector X' via the transition matrix:

$$X' = PX.$$

Given the matrix P, it is easy to calculate the probability of reaching any state from any other state in two steps. These probabilities are the entries of P^2. The reason is that the ijth entry of P^2 is

$$\sum_{k=1}^{m} p_{ik} p_{kj}.$$

In general, the ijth entry of P^n is the probability of reaching state j from state i in n steps. In terms of conditional probability, we write

$$p_{ij}^{(n)} = \Pr(X_{n+k} = j : X_k = i).$$

We denote by P^∞ the limit of P^n as $n \to \infty$ (if it exists). Thus, the ijth entry of P^∞ is $\lim_{n\to\infty} p_{ij}^{(n)}$.

A Markov chain is *regular* if there exists an n such that all the entries of P^n are positive. This means that for all states i and j, there is a positive probability of reaching state j from state i in n steps.

Theorem 11.1. Given a regular Markov chain with transition matrix P, the matrix P^∞ exists and its columns are each equal to X, where X is the unique probability vector satisfying the equation

$$X = PX.$$

For a proof of this theorem, see [Hel97].

We will show that the limiting probabilities exist in our case of the 3×3 transition matrix. The key is to find a way to exponentiate the matrix P. We do this by writing P in terms of a diagonal matrix D. Thus, we wish to write

$$P = ADA^{-1},$$

where D is a diagonal matrix. Then

$$P^n = AD^n A^{-1}.$$

The entries of D^n are easily computed, since D^n is a diagonal matrix whose diagonal entries are the entries of D raised to the nth power.

We find the matrices D and A using eigenvalues and eigenvectors. An *eigenvector* of a square matrix A is a nonzero vector v such that

$$Av = \lambda v,$$

for some scalar λ. We call λ the *eigenvalue* associated with v.

If $Av = \lambda v$, then $(A - \lambda I)v = 0$. The only way that this equation can have a nontrivial solution v (i.e., $v \neq 0$) is for the matrix $A - \lambda I$ to be singular. So λ is an eigenvalue of A if and only if $A - \lambda I$ is a singular matrix, that is, the determinant of this matrix is 0.

In our example, we want to solve the equation

$$\begin{vmatrix} -\lambda & 1/2 & 0 \\ 2/3 & -\lambda & 1 \\ 1/3 & 1/2 & -\lambda \end{vmatrix} = 0.$$

Working out the determinant, we obtain the *characteristic equation*[1]

$$\lambda^3 - \frac{5}{6}\lambda - \frac{1}{6} = 0.$$

---

[1]Our use of the term "characteristic equation" for a transition matrix is consistent with our use of the same term for a recurrence relation because a recurrence relation can be written via a transition matrix with the same characteristic equation.

The characteristic equation has solutions $\lambda_1 = 1$, $\lambda_2 = (-3 - \sqrt{3})/6$, and $(-3 + \sqrt{3})/6$. The eigenvector associated with the eigenvalue 1 is the vector $[3/13, 6/13, 4/13]$ that we found as the steady-state solution (we freely write vectors either horizontally or vertically). Note that we have normalized this vector so that the sum of the three coordinates is 1. What is the significance of the other two eigenvalues? They control the rate of convergence to the steady-state solution. Let's find eigenvectors corresponding to the other eigenvalues. We have

$$
\begin{bmatrix} 0 & 1/2 & 0 \\ 2/3 & 0 & 1 \\ 1/3 & 1/2 & 0 \end{bmatrix} \begin{bmatrix} x \\ y \\ z \end{bmatrix} = \frac{-3 - \sqrt{3}}{6} \begin{bmatrix} x \\ y \\ z \end{bmatrix}.
$$

From this system, we obtain the eigenvector $[\sqrt{3}, -1-\sqrt{3}, 1]$. Similarly, the eigenvalue $(-3 + \sqrt{3})/6$ corresponds to the eigenvector $[-\sqrt{3}, -1 - \sqrt{3}, 1]$.

Let v_1, v_2, and v_3 be the eigenvectors corresponding to the eigenvalues λ_1, λ_2, and λ_3, respectively. Let A be the 3×3 matrix whose columns are v_1, v_2, and v_3, i.e., $A = [v_1, v_2, v_3]$. Then, by the definition of eigenvalue and eigenvector, we obtain

$$
PA = A \begin{bmatrix} \lambda_1 & 0 & 0 \\ 0 & \lambda_2 & 0 \\ 0 & 0 & \lambda_3 \end{bmatrix}.
$$

It follows that

$$
P = ADA^{-1},
$$

where D is the diagonal matrix with diagonal entries λ_1, λ_2, and λ_3. This is the required decomposition of P.

Suppose that we start with the vector v_0. Let $A^{-1}v_0 = [\alpha_1, \alpha_2, \alpha_3]$. Then

$$
P^n v_0 = [v_1, v_2, v_3] \begin{bmatrix} 1 & 0 & 0 \\ 0 & \lambda_2^n & 0 \\ 0 & 0 & \lambda_3^n \end{bmatrix} \begin{bmatrix} \alpha_1 \\ \alpha_2 \\ \alpha_3 \end{bmatrix}
$$

$$
= \alpha_1 v_1 + \alpha_2 \lambda_2^n v_2 + \alpha_3 \lambda_3^n v_3.
$$

Since λ_2 and λ_3 are both less than 1 in absolute value, λ_2^n and λ_3^n tend to 0 as $n \to \infty$. Hence

$$
P^\infty v_0 = \alpha_1 v_1.
$$

Since $\alpha_1 v_1$ is a probability vector, $\alpha_1 = 1$. Therefore

$$
P^\infty v_0 = v_1.
$$

So the probability distribution tends to v_1, regardless of the starting probabilities.

Furthermore, since $PP^\infty = P^\infty$, each column of P^∞ is equal to v_1.

Example 11.2. Use a transition matrix to find a formula for the nth Fibonacci number, F_n. (This time the transition matrix does not represent probabilities.)

Solution: It is convenient to work with pairs of consecutive Fibonacci numbers. Thus, the Fibonacci recurrence yields

$$\begin{bmatrix} F_{n+1} \\ F_n \end{bmatrix} = \begin{bmatrix} 1 & 1 \\ 1 & 0 \end{bmatrix} \begin{bmatrix} F_n \\ F_{n-1} \end{bmatrix}, \quad n \geq 1.$$

Hence

$$\begin{bmatrix} F_{n+1} \\ F_n \end{bmatrix} = \begin{bmatrix} 1 & 1 \\ 1 & 0 \end{bmatrix}^n \begin{bmatrix} 1 \\ 0 \end{bmatrix}, \quad n \geq 0.$$

Let

$$M = \begin{bmatrix} 1 & 1 \\ 1 & 0 \end{bmatrix}.$$

If we can write M as

$$M = ADA^{-1},$$

then it will follow that

$$M^n = AD^n A^{-1}.$$

If D is a diagonal matrix, then D^n is easily computed, so in this case we would have a simple way of computing the nth Fibonacci number directly. If

$$M = \begin{bmatrix} v_1 & v_2 \end{bmatrix} \begin{bmatrix} \lambda_1 & 0 \\ 0 & \lambda_2 \end{bmatrix} \begin{bmatrix} v_1 & v_2 \end{bmatrix}^{-1},$$

then

$$M \begin{bmatrix} v_1 & v_2 \end{bmatrix} = \begin{bmatrix} v_1 & v_2 \end{bmatrix} \begin{bmatrix} \lambda_1 & 0 \\ 0 & \lambda_2 \end{bmatrix},$$

and we see that v_1 and v_2 are eigenvectors with eigenvalues λ_1 and λ_2, respectively. The eigenvalues are

$$\lambda_1 = \frac{1 + \sqrt{5}}{2}, \quad \lambda_2 = \frac{1 - \sqrt{5}}{2},$$

with corresponding eigenvectors

$$v_1 = \begin{bmatrix} \lambda_1 \\ 1 \end{bmatrix}, \quad v_2 = \begin{bmatrix} \lambda_2 \\ 1 \end{bmatrix}.$$

Now we have

$$\begin{bmatrix} F_{n+1} \\ F_n \end{bmatrix} = \begin{bmatrix} \lambda_1 & \lambda_2 \\ 1 & 1 \end{bmatrix} \begin{bmatrix} \lambda_1^n & 0 \\ 0 & \lambda_2^n \end{bmatrix} \begin{bmatrix} \lambda_1 & \lambda_2 \\ 1 & 1 \end{bmatrix}^{-1} \begin{bmatrix} 1 \\ 0 \end{bmatrix}.$$

Since

$$\begin{bmatrix} \lambda_1 & \lambda_2 \\ 1 & 1 \end{bmatrix}^{-1} = \frac{1}{\lambda_1 - \lambda_2} \begin{bmatrix} 1 & -\lambda_2 \\ -1 & \lambda_1 \end{bmatrix},$$

we obtain

$$F_n = \frac{\lambda_1^n - \lambda_2^n}{\lambda_1 - \lambda_2}$$

$$= \frac{1}{\sqrt{5}} \left(\frac{1 + \sqrt{5}}{2} \right)^n - \frac{1}{\sqrt{5}} \left(\frac{1 - \sqrt{5}}{2} \right)^n, \quad n \geq 0.$$

■

In general, suppose that the sequence $\{a_n\}$ satisfies a linear recurrence relation with constant coefficients, i.e., a relation

$$a_n = c_1 a_{n-1} + c_2 a_{n-2} + \cdots + c_k a_{n-k},$$

with constants $c_1, \ldots, c_k$, for all $n \geq k$.

In terms of a transition matrix, we have

$$\begin{bmatrix} a_n \\ a_{n-1} \\ \vdots \\ a_{n-k+2} \\ a_{n-k+1} \end{bmatrix} = \begin{bmatrix} c_1 & c_2 & \cdots & c_{k-1} & c_k \\ 1 & 0 & \cdots & 0 & 0 \\ 0 & 1 & \cdots & 0 & 0 \\ \vdots & & \ddots & & \vdots \\ 0 & & \cdots & 1 & 0 \end{bmatrix} \begin{bmatrix} a_{n-1} \\ a_{n-2} \\ \vdots \\ a_{n-k+1} \\ a_{n-k} \end{bmatrix}, \quad n \geq k.$$

Call the $k \times k$ transition matrix C. The characteristic polynomial of C is

$$\det(C - xI) = \begin{vmatrix} c_1 - x & c_2 & c_3 & \cdots & c_k \\ 1 & -x & 0 & \cdots & 0 \\ 0 & 1 & -x & \cdots & 0 \\ \vdots & & \ddots & \ddots & \vdots \\ 0 & & \cdots & 1 & -x \end{vmatrix}.$$

The value of this determinant is

$$(c_1 - x)(-x)^{k-1} - 1 \cdot (c_2(-x)^{k-2} - 1 \cdot (c_3(-x)^{k-3} - \cdots))$$

$$= (-1)^k (x^k - x^{k-1}c_1 - x^{k-2}c_2 - x^{k-3}c_3 - \cdots - c_k).$$

This is $(-1)^k$ times the characteristic polynomial for the recurrence relation. Suppose that the roots of this polynomial are $r_1, \ldots, r_k$, with the r_i distinct. Then C is diagonalizable as

$$C = ADA^{-1},$$

where D is a diagonal matrix with diagonal entries $r_1, \ldots, r_k$. Hence

$$
\begin{bmatrix} a_n \\ a_{n-1} \\ \vdots \\ a_{n-k+1} \end{bmatrix} = AD^n A^{-1} \begin{bmatrix} a_{k-1} \\ a_{k-2} \\ \vdots \\ a_0 \end{bmatrix}.
$$

The top relation of this system is

$$
a_n = \alpha_1 r_1^n + \cdots + \alpha_k r_k^n, \quad n \geq 0,
$$

where $\alpha_1, \alpha_2, \ldots, \alpha_k$ are constants.

What is the geometric significance of eigenvalues and eigenvectors? Let's take a look at the Fibonacci numbers transition matrix:

$$
M = \begin{bmatrix} 1 & 1 \\ 1 & 0 \end{bmatrix}
$$

This matrix represents an action on the real plane $\mathbf{R}^2$. That is to say, each vector $v \in \mathbf{R}^2$, upon multiplication by M, is transformed into a new vector $v' \in \mathbf{R}^2$.

We note that the zero vector is mapped to itself under this action. Also,

$$
\begin{bmatrix} 1 & 1 \\ 1 & 0 \end{bmatrix} \begin{bmatrix} 1 \\ 0 \end{bmatrix} = \begin{bmatrix} 1 \\ 1 \end{bmatrix}, \qquad \begin{bmatrix} 1 & 1 \\ 1 & 0 \end{bmatrix} \begin{bmatrix} 0 \\ 1 \end{bmatrix} = \begin{bmatrix} 1 \\ 0 \end{bmatrix}.
$$

Furthermore, the eigenvalues and eigenvectors are defined so that $Mv_1 = \lambda_1 v_1$ and $Mv_2 = \lambda_2 v_2$. Figure 11.1 shows the action. We can see in the figure that the vectors v_1 and v_2 are linearly independent; hence they form a coordinate system for the plane. Given any $v \in \mathbf{R}^2$, we can write

$$
v = av_1 + bv_2.
$$

The "coordinates" of v with respect to the new basis are a and b. Now it is easy to see what happens to v under the transformation given by M. We have

$$
\begin{aligned}
Mv &= M(av_1 + bv_2) \\
&= aMv_1 + bMv_2 \\
&= a\lambda_1 v_1 + b\lambda_2 v_2.
\end{aligned}
$$

Therefore, the vector with coordinates (a, b) is sent to a new vector $v' = Mv$ with coordinates $(a\lambda_1, b\lambda_2)$. Simply put, the coordinates of v are multiplied by the eigenvalues of M.

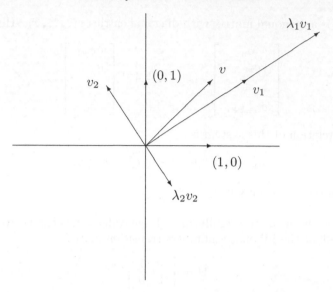

FIGURE 11.1: The action of the matrix M.

Exercises

1. What is the steady-state solution of the Markov chain with transition matrix

$$P = \begin{bmatrix} 0 & 1/3 & 1/2 \\ 1/4 & 0 & 1/2 \\ 3/4 & 2/3 & 0 \end{bmatrix} ?$$

What are the eigenvalues and eigenvectors of this matrix? Calculate P^∞.

◇2. Suppose that the Markov chain of the previous exercise starts with the probability distribution $[1, 0, 0]$. After 100 steps, how close is the probability distribution to the steady-state solution?

3. Does the transition matrix

$$\begin{bmatrix} 0 & 1 & 0 \\ 0 & 0 & 1 \\ 1 & 0 & 0 \end{bmatrix}$$

represent a regular Markov chain?

4. Let
$$M = \begin{bmatrix} -15 & -2 \\ 153 & 20 \end{bmatrix}.$$

Find M raised to the 100th power.

5. Find a 2×2 non-identity matrix M such that $M^5 = I_2$.

6. Find the 2×2 matrix that reflects each vector in the plane with respect to the x-axis. What are the eigenvalues and eigenvectors of this matrix?

7. Find the 2×2 matrix that reflects each vector in the plane with respect to the line $y = x + 1$. What are the eigenvalues and eigenvectors of this matrix?

8. Find the 2×2 matrix that projects each vector in the plane orthogonally onto the line $y = 3x$. What are the eigenvalues and eigenvectors of this matrix?

9. Use matrices to find a formula for a_n, where $a_0 = 0$, $a_1 = 1$, and $a_n = 2a_{n-1} - a_{n-2}$, $n \geq 2$.

10. Use matrices to find a formula for b_n, where $b_0 = 3$, $b_1 = 10$, $b_3 = 38$, and $b_n = 10b_{n-1} - 31b_{n-2} + 30b_{n-3}$, for $n \geq 3$.

11. Find a 2×2 matrix M with integer entries such that $M^2 - M - I = 0$.

12. A group of n people, $X_1, \ldots, X_n$, are throwing a ball back and forth. Suppose that at each increment of time, the person currently holding the ball throws it to one of the others. Let the steady-state probability distribution be $[p_1, p_2, \ldots, p_n]$; that is, p_i, for $1 \leq i \leq n$, is the probability that X_i has the ball at any given time.

    (a) Show that $p_1 = 1/n$ regardless of who X_1 throws the ball to (as long as the others throw the ball to everyone with equal likelihood).

    (b) Suppose that everyone throws the ball to everyone else with equal likelihood except that X_1 always throws the ball to X_2. Show that
    $$[p_1, p_2, \ldots, p_n] = \left[\frac{1}{n}, \frac{2n-2}{n^2}, \frac{n-1}{n^2}, \ldots, \frac{n-1}{n^2} \right].$$

    Thus, on average, X_2 has the ball twice as often as every other person except X_1, who has the ball more often than every other person except X_2.

Chapter 12

Random Tournaments

A large random tournament almost assuredly has the property that for every 10 teams, there is a team that beats all 10.

In a large random tournament, almost assuredly every vertex is both a King and a Serf.

A tournament is a complete graph (we will see more about graphs later in the book) in which each edge is replaced by an arrow (i.e., a directed edge). If an edge goes from vertex v to vertex w, say that team v beats team w in a game.

A tournament has *Property k* if, for every k vertices, some vertex beats all of them. For example, the tournament of Figure 12.1 has Property 1.

Figure 12.2 shows a tournament with Property 2. The tournament is called a quadratic residue tournament. The vertices are labeled 0 through 7. The assignment of the direction of the edges is based on modulo 7 arithmetic. Vertex i is directed to vertex j if $j - i$ is a square modulo 7. The squares modulo 7 are $0^2 \equiv 0$, $(\pm 1)^2 \equiv 1$, $(\pm 2)^2 \equiv 4$, and $(\pm 3)^2 \equiv 2$. Since a nonzero element x is a square modulo 7 if and only if $-x$ is a non-square modulo 7, the direction of edges is well-defined.

It isn't obvious how to construct a tournament with, say, Property 10. However, we can prove that such a tournament exists without constructing

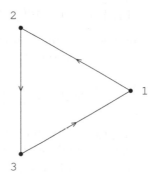

FIGURE 12.1: A tournament with Property 1.

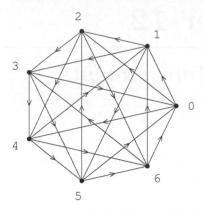

FIGURE 12.2: A tournament with Property 2.

it. The main result is sometimes called Schütte's theorem but was proved by Paul Erdős (1913–1996). The proof uses probability.

Theorem 12.1 (P. Erdős, 1963). Let k be a positive integer. For some integer n, there exists a tournament on n vertices with Property k.

Proof. Let k be a positive integer and n an integer to be determined later. Let the edges of the tournament on n vertices be directed one way or the other at random, with equal probability, and independently of all the other edges. We call such a tournament a "random tournament." For every subset S of k vertices, denote by A_S be the event that there is no vertex in the tournament that beats all the vertices of S. In order for A_S to occur, none of the $n - k$ vertices in the complement of S may be directed to all vertices of S. Hence

$$\Pr(A_S) = (1 - 2^{-k})^{n-k}.$$

Since the probability of a union of events is at most equal to the sum of the probabilities of the individual events (the case $t = 1$ in Theorem 10.14), we have

$$\Pr\left(\bigcup_S A_S\right) \le \sum_S \Pr(A_S) = \binom{n}{k}(1 - 2^{-k})^{n-k}.$$

Note that $\Pr(\bigcup_S A_S)$ is the probability that the tournament does not have Property k. We have an upper bound for this union that is the product of two terms: $\binom{n}{k}$ and $\left(1 - 2^{-k}\right)^{n-k}$. Since k is fixed, $\binom{n}{k}$ is a polynomial in n (of degree k), while $\left(1 - 2^{-k}\right)^{n-k}$ is an exponential function in n (with a base less than 1). As n tends to infinity, the exponential function dominates and the product of the two terms tends to 0. Since the upper bound tends to 0, the probability of the union must be less than 1 for some n. By taking complements, we see that for such an n, a random tournament on n vertices has Property k with positive probability. This probability is equal to the number of tournaments with Property k divided by the total number of tournaments on n vertices. Therefore, there exists a tournament on n vertices with Property k. ∎

Following the method of the proof, we deduce that there exists a tournament with Property 10 on 102,653 vertices, since

$$\binom{n}{10}(1 - 2^{-10})^{n-10} < 1,$$

for $n = 102653$.

Now we consider another type of question about tournaments.

In a tournament, a *King* is a vertex from which every other vertex can be reached in one or two directed steps. A *Serf* is a vertex that can be reached from every other vertex in one or two directed steps. Every vertex in the tournaments of Figures 12.1 and 12.2 is both a King and a Serf. We will prove that this situation is typical.

The *outdegree* of a vertex is the number of edges directed away from that vertex. The *indegree* of a vertex is the number of edges directed to it.

Theorem 12.2 (H. G. Landau, 1951). Every tournament has a King.

Proof. Consider a vertex v of maximum outdegree. We will prove that v is a King of the tournament. Suppose that there are edges directed from v to r vertices, $u_1, \ldots, u_r$. Assume that there is a vertex w that cannot be reached in one or two steps from v. Then w is not among the u_i and there are edges directed from w to all the u_i and to v. But then the outdegree of w is at least $r + 1$, contradicting the choice of v. ∎

Theorem 12.3 (Stephen B. Maurer, 1980). The probability that every vertex in a random tournament on n vertices is both a King and a Serf tends to 1 as n tends to infinity.

Proof. We use the probabilistic method. A tournament lacks the desired property if and only if there exists a pair of vertices v and w with $v \to w$ such that there is no directed path of length 2 from w to v. This happens with probability at most

$$\binom{n}{2}\left(\frac{3}{4}\right)^{n-2} \longrightarrow 0 \quad (\text{as } n \to \infty).$$

∎

Exercises

◇1. Find a value of n for which there is a tournament on n vertices with Property 3. Can you construct a tournament with Property 3?

◇2. Find a value of n for which there is a tournament on n vertices with Property 12.

◇3. Find a value of n for which a random tournament on n vertices has Property 12 with probability greater than 0.5.

⋆4. Prove that the tournament on seven vertices with Property 2 shown in Figure 12.2 is unique up to isomorphism.

†5. Prove that no tournament on $2^n - 1$ vertices has Property n.

6. Prove that every tournament has a directed path that visits each vertex exactly once.

7. Find a formula for the maximum number of directed 3-cycles in a tournament of order n.

8. Prove that every tournament has a Serf.

9. Construct a quadratic residue tournament on 11 vertices. How many Kings and Serfs does it have?

10. A vertex which reaches every other vertex in one step is called an *Emperor*. Prove that a tournament with no Emperor has at least three Kings.

†11. Show that a tournament on $n > 4$ vertices can have any number of Kings between 1 and n except 2. What happens when $n = 3$ or 4?

12. In the proof of Theorem 12.2, we tacitly assumed that the tournament has a finite number of vertices. Is it true that every tournament on a countably infinite set of vertices has a King?

Part V

Number Theory

Chapter 13

Divisibility of Factorials and Binomial Coefficients

In any row of Pascal's triangle, any two numbers aside from the 1s have a common factor.

Many observations can be made about the number-theoretic properties of factorials and binomial coefficients.

Example 13.1. How many 0s occur at the right of 40!?

Solution: The 0s at the right of 40! appear due to factors of 2 and 5 among the numbers 1, 2, ..., 40. Since there are more 2s than 5s, the number of 0s is determined by the exponent of 5 that divides 40!. This number is

$$\sum_{k=1}^{\infty} \left\lfloor \frac{40}{5^k} \right\rfloor = \left\lfloor \frac{40}{5} \right\rfloor + \left\lfloor \frac{40}{25} \right\rfloor = 8 + 1 = 9.$$

∎

In the following discussion, let p be a prime number.

Proposition 13.2. The power to which a prime p divides $n!$ is given by

$$\sum_{k=1}^{\infty} \left\lfloor \frac{n}{p^k} \right\rfloor.$$

This series is actually a finite one, since $p^k > n$ for k sufficiently large. Denote by $d_b(n)$ the sum of the "digits" in the base-b representation of n. For instance, if the base-3 representation of n is 102012, then $d_3(n) = 6$.

Proposition 13.3. The power of 2 that divides $n!$ is $n - d_2(n)$.

Proof. Let the base-2 representation of n be

$$n = b_k b_{k-1} \ldots b_1 b_0.$$

Then $n = \sum_{i=0}^{k} b_i 2^i$, so that the exponent of 2 that divides $n!$ is

$$
\begin{aligned}
\sum_{i=1}^{\infty} \left\lfloor \frac{n}{2^i} \right\rfloor &= (b_1 + 2b_2 + 2^2 b_3 + \cdots + 2^{k-1} b_k) \\
&\quad + (b_2 + 2b_3 + \cdots + 2^{k-2} b_k) \\
&\quad + \cdots \\
&\quad + b_k \\
&= (2^0 - 1)b_0 + (2^1 - 1)b_1 + (2^2 - 1)b_2 + \cdots + (2^k - 1)b_k \\
&= n - d_2(n).
\end{aligned}
$$

∎

The previous result is a special case of a theorem of Adrien-Marie Legendre (1752–1833).

Theorem 13.4 (A.-M. Legendre, 1808). The exponent of p that divides $n!$ is
$$
\frac{n - d_p(n)}{p - 1}.
$$

Next, we turn to the question of the divisibility of binomial coefficients by primes.

Proposition 13.5. If p is a prime and $1 \le k \le p - 1$, then $\binom{p}{k}$ is divisible by p.

Proof. The numerator of $p!/(k!(p-k)!)$ is a multiple of p and p does not divide the denominator. ∎

The following surprising (and delightful) result was discovered by Ernst Kummer (1810–1893).

Theorem 13.6 (E. Kummer, 1852). The exponent to which a prime p divides the binomial coefficient $\binom{n}{k}$ is equal to the number of "carries" when k and $n - k$ are added in base p.

Proof. We will prove the result in the base 2 case. Let $j = n - k$. The exponent to which 2 divides $\binom{n}{k}$ is

$$
n - d_2(n) - (j - d_2(j) + k - d_2(k)) = d_2(j) + d_2(k) - d_2(n).
$$

Assume that the binary representation of n requires l binary digits. For $1 \le i \le l$, let n_i, j_i, and k_i be the ith binary digit of the expansion of n, j, and k, respectively; let $c_i = 1$ if there is a carry in the ith place when j

and k are added (in binary), and $c_i = 0$ if there is no carry. Also, define $c_{-1} = 0$. From the definition of "carry," $n_i = j_i + k_i + c_{i-1} - 2c_i$, for $1 \leq i \leq l$. Hence, the exponent to which 2 divides $\binom{n}{k}$ is

$$\sum_{i=0}^{l}(j_i + k_i - n_i) = \sum_{i=0}^{l}(2c_i - c_{i-1}) = \sum_{i=0}^{l} c_i.$$

∎

Corollary 13.7. For $e \geq 1$ and $1 \leq m < p^e$, we have

$$\binom{p^e}{m} \equiv 0 \pmod{p}.$$

Now we hear from another great mathematician, François Édouard Anatole Lucas (1842–1891). Lucas' theorem gives a practical method for calculating $\binom{a}{b} \bmod p$.

Theorem 13.8 (E. Lucas, 1878). Let $0 \leq a_i, b_i < p$, for $1 \leq i \leq k$. Then

$$\binom{a_0 + a_1 p + a_2 p^2 + \cdots + a_k p^k}{b_0 + b_1 p + b_2 p^2 + \cdots + b_k p^k} \equiv \binom{a_0}{b_0}\binom{a_1}{b_1}\binom{a_2}{b_2}\cdots\binom{a_k}{b_k} \pmod{p}.$$

Proof. The left side counts the ways of choosing $b_0 + b_1 p + b_2 p^2 + \cdots + b_k p^k$ balls from a set of $a_0 + a_1 p + a_2 p^2 + \cdots + a_k p^k$ balls. Suppose that the balls to be selected are in boxes, with a_0 boxes containing a single ball each, a_1 boxes containing p balls each, a_2 boxes containing p^2 balls each, $\ldots$, and a_k boxes containing p^k balls each. In selecting the balls from boxes, any choice of some but not all the balls from a box leads to a contribution of 0 $(\bmod\ p)$, since $\binom{p^e}{m} \equiv 0 \pmod{p}$, for $1 \leq m < p^e$. Hence, the only selections that matter (modulo p) are those that take none or all the balls from a particular box. This means that we need to select b_i boxes from a set of a_i boxes from which to take all the balls, for $0 \leq i \leq k$. The number of ways to do this is

$$\binom{a_0}{b_0}\binom{a_1}{b_1}\binom{a_2}{b_2}\cdots\binom{a_k}{b_k}.$$

∎

We say that the base-p representation of n "dominates" the base-p representation of k if the number in each place of the base-p representation of n is at least equal to the number in the corresponding place in the base-p representation of k.

Corollary 13.9. The binomial coefficient $\binom{n}{k}$ is divisible by p if and only if the base-p representation of n does not dominate the base-p representation of k.

Example 13.10. Is $\binom{59}{11}$ divisible by 7?

Solution: Let's see: $59 = 1 \cdot 7^2 + 1 \cdot 7 + 3$ and $11 = 1 \cdot 7 + 4$. Since the base-7 representation of 59 does not dominate the base-7 representation of 11, we know that $\binom{59}{11}$ is divisible by 7. ∎

We conclude with a discussion of the chapter teaser about numbers in a row of Pascal's triangle, a result due to Paul Erdős (1913–1996) and George Szekeres (1911–2005).

Theorem 13.11 (P. Erdős and G. Szekeres, 1978)**.** In any row of Pascal's triangle, any two numbers aside from the 1s have a common factor.

For example, in the sixth row,

$$1 \quad 6 \quad 15 \quad 20 \quad 15 \quad 6 \quad 1,$$

the entries 6 and 15 have a common factor 3, while 6 and 20 have a common factor 2, and 15 and 20 have a common factor 5.

Proof. Suppose that the numbers are $\binom{n}{j}$ and $\binom{n}{k}$, with $0 < j < k < n$. Then, by the subcommittee identity (Proposition 3.3),

$$\binom{n}{k}\binom{k}{j} = \binom{n}{j}\binom{n-j}{k-j}.$$

Since $\binom{n}{j}$ divides the right side of this equation, it also divides the left side. If $\binom{n}{j}$ and $\binom{n}{k}$ have no common factor, then $\binom{n}{j}$ divides $\binom{k}{j}$, but this is impossible since $\binom{n}{j} > \binom{k}{j}$. ∎

Exercises

1. How many 0s occur at the right of 1000!?

2. Find the smallest integer n such that $n!$ ends with exactly one hundred 0s.

3. To what power does 2 divide $(2^{100} + 1)!$?

4. To what power does 2 divide $(2^{100} - 1)!$?

5. Does 7 divide $\binom{7^{50}+6\cdot7^{25}+6}{3\cdot7^{25}+1}$?

6. Show that $(kn)!$ is divisible by $(n!)^k$, for any positive integers k and n.

7. Prove that $\binom{n}{k}$ is divisible by n if $\gcd(n, k) = 1$.

8. Let k and m be integers such that $0 \le m \le 2^k - 1$. Prove that the binomial coefficient $\binom{2^k-1}{m}$ is odd.

9. Investigate the entries of Pascal's triangle modulo 2. What pattern do you find?

10. Prove that in any row of Pascal's triangle, the number of odd numbers is a power of 2.

11. Is $\binom{2^{10}+2^5+1}{2^5+1}$ divisible by 2?

12. Use Lucas' theorem to calculate $\binom{7^{100}+7^3+2\cdot7+5}{7^3+4}$ mod 7.

◇13. Paul Erdős proved that there is only one nontrivial perfect power (not a first power) of the form $\binom{n}{k}$, with $3 \le k \le n - 3$. Use a computer to find this binomial coefficient.

14. Show that $n!$ cannot be a perfect square greater than 1.

15. Notice that $6! = 3!5!$. Can you find other instances of integers a, b, and c, all greater than 1, such that $a!b! = c!$? Is there any pattern to these numbers?

16. Prove the following result of Erdős and Szekeres:

$$\gcd\left(\binom{n}{i}, \binom{n}{j}\right) \ge 2^i,$$

where $0 < i \le j \le n/2$.

†◇17. Use a computer to find the only two ordered pairs (n, k), with $1 < k < n/2$ for which

$$\sum_{k=0}^{e} \binom{n}{k}$$

is a power of 2.

Note. The values (n, k) are the feasible parameters for perfect binary codes.

Chapter 14

Covering Systems

There exists an odd integer k such that $k + 2^n$ is a composite number for every positive integer n.

There exists a sequence of composite numbers satisfying the Fibonacci recurrence relation with relatively prime initial values.

A *covering system* is a collection of congruences of the form $x \equiv a_i$ (mod m_i), for $1 \leq i \leq k$, where the m_i are integers greater than 1, such that every integer x satisfies at least one of the congruences. An example of a covering system is the set of congruences

$$x \equiv 0 \pmod 2$$

$$x \equiv 0 \pmod 3$$

$$x \equiv 1 \pmod 4$$

$$x \equiv 1 \pmod 6$$

$$x \equiv 11 \pmod{12}.$$

It's easy to verify that these congruences are a covering system. Write the integers 0 through 11 and cross out the integers covered by each congruence. Every integer will be crossed out.

We will show that there exists an odd integer k such that $k + 2^n$ is prime for no positive integer n. The method of proof, due to Paul Erdős, uses a covering system.

The value $k = 23$ does not satisfy the requirement of the problem but let's see what happens with this value. We have $23 + 2^1 = 25$, which is divisible by 5. Hence, $23 + 2^n$ is divisible by 5 if $2^n \equiv 2 \pmod 5$. The powers of 2 modulo 5 form the cycle $\{2, 4, 3, 1\}$. It follows that $23 + 2^n$ is composite (divisible by 5) for $n \equiv 1 \pmod 4$. Similarly, $23 + 2^2 = 27$, which is divisible by 3. Hence, $23 + 2^n$ is divisible by 3 if $2^n \equiv 1 \pmod 3$. The powers of 2 modulo 3 form the cycle $\{2, 1\}$. It follows that $k + 2^n$ is composite (divisible by 3) for $n \equiv 0 \pmod 2$. We have ruled out two infinite arithmetic progressions as choices for k, namely, all solutions to

$n \equiv 1 \pmod 4$ and $n \equiv 0 \pmod 2$. The smallest positive integer not ruled out is 3, and $23 + 2^3 = 31$, a prime.

Erdős' method is to choose a set of primes to rule out all possible values of n. We see from the example $k = 23$ that we should investigate cycles of powers of 2 modulo various primes. The following table shows our selection of primes and the length of each corresponding cycle of powers of 2.

prime	length of cycle of powers of 2
3	2
5	4
7	3
13	12
17	8
241	24

We form a covering system with the lengths as moduli:

$$x \equiv 1 \pmod 2$$

$$x \equiv 0 \pmod 4$$

$$x \equiv 0 \pmod 3$$

$$x \equiv 2 \pmod{12}$$

$$x \equiv 2 \pmod 8$$

$$x \equiv 22 \pmod{24}.$$

To find an odd integer k such that $k + 2^n$ is never prime, we look for a solution to the system of congruences

$$k \equiv 1 \pmod 2$$

$$k \equiv -2^1 \pmod 3$$

$$k \equiv -2^0 \pmod 5$$

$$k \equiv -2^0 \pmod 7$$

$$k \equiv -2^2 \pmod{13}$$

$$k \equiv -2^2 \pmod{17}$$

$$k \equiv -2^{22} \pmod{241}.$$

(The purpose of the first congruence is to ensure that k is odd.) This will furnish an arithmetic progression of integers k that satisfy the condition of the problem. For instance, if $n \equiv 0 \pmod 4$, then $2^n \equiv 1 \pmod 5$ and since $k \equiv -1 \pmod 5$, it follows that $k + 2^n$ is divisible by 5 (and hence composite).

We find a simultaneous solution to the above congruences using the Chinese remainder theorem.[1]

Theorem 14.1 (Chinese Remainder Theorem). If $n_1, n_2, \ldots, n_k$ are pairwise relatively prime numbers, and $r_1, r_2, \ldots, r_k$ are any integers, then there exists an integer x satisfying the simultaneous congruences

$$x \equiv r_1 \pmod{n_1}$$

$$x \equiv r_2 \pmod{n_2}$$

$$\vdots$$

$$x \equiv r_k \pmod{n_k}.$$

Furthermore, x is unique modulo $n_1 n_2 \ldots n_k$.

The solution to our problem will illustrate the constructive nature of the proof of the Chinese remainder theorem.

Let $m = 2 \cdot 3 \cdot 5 \cdot 7 \cdot 13 \cdot 17 \cdot 241 = 11184810$. We solve the following system of congruences:

$$(m/2)k_1 \equiv 1 \pmod 2$$

$$(m/3)k_2 \equiv -2^1 \pmod 3$$

$$(m/5)k_3 \equiv -2^0 \pmod 5$$

$$(m/7)k_4 \equiv -2^0 \pmod 7$$

$$(m/13)k_5 \equiv -2^2 \pmod{13}$$

$$(m/17)k_6 \equiv -2^2 \pmod{17}$$

$$(m/241)k_7 \equiv -2^{22} \pmod{241}.$$

Choosing the values $k_1 = 1$, $k_2 = 2$, $k_3 = 2$, $k_4 = 2$, $k_5 = 12$, $k_6 = 1$, and $k_7 = 210$ yields the solution

$$k \equiv \left(\frac{m}{2}\right)1 + \left(\frac{m}{3}\right)2 + \left(\frac{m}{5}\right)2 + \left(\frac{m}{7}\right)2 + \left(\frac{m}{13}\right)12 + \left(\frac{m}{17}\right)1 + \left(\frac{m}{241}\right)210$$

$$\equiv 41446999 \pmod m.$$

---

[1] The Chinese remainder theorem first appeared around 400 in the book *Sun Tzu Suan Ching* ("Sun Tzu's Calculation Classic") by Sun Tzu.

Take k to be the smallest positive integer in this congruence class, i.e.,

$$k = 41446999 - 3 \cdot 11184810 = 7892569.$$

The number $7892569 + 2^n$ is a prime for no positive integer n.

Let's turn to another problem that can be solved with a covering system, that of constructing a Fibonacci-like sequence of composite numbers.

It is not known whether there are infinitely many prime Fibonacci numbers. Certainly, there are infinitely many composite Fibonacci numbers, since every third Fibonacci number is even. We will show that there exist relatively prime positive integers a and b, such that the Fibonacci-like sequence defined by the Fibonacci recurrence relation and the initial values a and b contains *no* prime numbers. The first example of such a sequence was found in 1990 by Donald Knuth. We will show an example discovered in 2004 by Maxim Vsemirnov.

Our sequence $\{a_n\}$ is defined by

$$a_0 = a, \ a_1 = b, \ a_n = a_{n-1} + a_{n-2}, \ n \geq 2.$$

It follows by mathematical induction that

$$a_n = aF_{n-1} + bF_n, \quad n \geq 1.$$

We define 17 quadruples of integers (p_i, m_i, r_i, c_i), where $1 \leq i \leq 17$. These quadruples satisfy the following properties:
(1) each p_i is prime;
(2) $p_i \mid F_{m_i}$;
(3) the congruences $x \equiv r_i \pmod{m_i}$ cover all the integers; that is, given any integer n, at least one of the congruences is satisfied by n.
We define

$$a \equiv c_i F_{m_i - r_i} \pmod{p_i}, \quad b \equiv c_i F_{m_i - r_i + 1} \pmod{p_i}, \quad 1 \leq i \leq 17.$$

Such integers a and b exist by (1) and the Chinese remainder theorem.

Now, using the identity from Exercise 6 of Chapter 3, we have

$$a_n \equiv c_i F_{m_i - r_i} F_{n-1} + c_i F_{m_i - r_i + 1} F_n \pmod{p_i}$$

$$\equiv c_i (F_{m_i - r_i} F_{n-1} + F_{m_i - r_i + 1} F_n) \pmod{p_i}$$

$$\equiv c_i F_{m_i - r_i + n} \pmod{p_i}.$$

From (2) and the fact that $F_m \mid F_{mn}$ (exercise), we have $p_i \mid a_n$, where i is given by property (3).

The following collection of 17 quadruples is found by computer experimentation. (The values of c_i help to keep the solutions a and b small.)

$(3, 4, 3, 2)$	$(2, 3, 1, 1)$	$(5, 5, 4, 2)$
$(7, 8, 5, 3)$	$(17, 9, 2, 5)$	$(11, 10, 6, 6)$
$(47, 16, 9, 34)$	$(19, 18, 14, 14)$	$(61, 15, 12, 29)$
$(23, 24, 17, 6)$	$(107, 36, 8, 19)$	$(31, 30, 0, 21)$
$(1103, 48, 33, 9)$	$(181, 90, 80, 58)$	$(41, 20, 18, 11)$
	$(541, 90, 62, 85)$	$(2521, 60, 48, 306)$

Using the (constructive) Chinese remainder theorem, we find

$$a = 106276436867, \ b = 35256392432.$$

The smallest initial values that generate a Fibonacci-like sequence of composite numbers are not known.

Exercises

1. Give an example of a covering system with three congruences having distinct moduli.

$\diamond$2. Find a smaller odd positive integer than $k = 7892569$ for which $k + 2^n$ is a composite number for every positive integer n.

3. Prove that there are infinitely many odd positive integers that are not equal to the sum of a prime number and a power of 2.

4. Show that there exists an integer k, not divisible by 3, such that $k + 3^n$ is prime for no positive integer n.

†5. Prove the Chinese remainder theorem.

†⋆6. An *exact covering system* is a covering system in which each integer satisfies exactly one of the congruences. Prove that there does not exist an exact covering system for the integers with distinct moduli.

†7. Show that no Fibonacci number F_n with n odd has a prime factor of the form $4k + 3$.

†8. Prove that $F_m \mid F_n$ if and only if $m \mid n$.

†9. Prove that $\gcd(F_m, F_n) = F_{\gcd(m,n)}$.

10. Check that the congruences $x \equiv r_i \pmod{m_i}$ given by the 17 quadruples cover all integers.

Chapter 15

Partitions of an Integer

There are 24,061,467,864,032,622,473,692,149,727,991 partitions of 1000 as unordered sums of positive integers.

A *partition* of a positive integer n is a collection of positive integers (order unimportant) whose sum is n. The integers in a partition are called *parts*. We denote by $p(n)$ the number of partitions of n, and by $p(n,k)$ the number of partitions of n into exactly k parts. Clearly,

$$p(n) = \sum_{k=1}^{n} p(n,k).$$

We call $p(n)$ and $p(n,k)$ *partition numbers*.

Example 15.1. Determine $p(5,k)$, for $1 \leq k \leq 5$, and $p(5)$.

Solution:

$$p(5,1) = 1 \quad (5)$$
$$p(5,2) = 2 \quad (4+1,\ 3+2)$$
$$p(5,3) = 2 \quad (3+1+1,\ 2+2+1)$$
$$p(5,4) = 1 \quad (2+1+1+1)$$
$$p(5,5) = 1 \quad (1+1+1+1+1)$$

$$p(5) = p(5,1) + p(5,2) + p(5,3) + p(5,4) + p(5,4)$$
$$= 1 + 2 + 2 + 1 + 1 = 7$$

■

As in our example, we typically represent a partition of n with the parts in monotonically decreasing order:

$$n = \lambda_1 + \lambda_2 + \cdots + \lambda_k, \quad \lambda_1 \geq \lambda_2 \geq \cdots \geq \lambda_k.$$

Recall from Exercise 28 of Chapter 3 that a composition of n is an ordered sum of positive integers equal to n. The number of compositions of n is 2^{n-1}. In a partition, the order of the summands is unimportant. A formula for the number of partitions of n is known but complicated. In 1918 G. H. Hardy (1877–1947) and Srinivasa Ramanujan (1887–1920) found the asymptotic formula

$$p(n) \sim \frac{e^{\pi\sqrt{2n/3}}}{4n\sqrt{3}}.$$

We will discuss recurrence relations for partition numbers and special properties of certain restricted types of partition numbers.

We can calculate $p(n, k)$ and $p(n)$ via a simple recurrence formula:

$$p(1, 1) = 1,$$

$$p(n, k) = 0, \quad k > n \text{ or } k = 0,$$

$$p(n, k) = p(n - 1, k - 1) + p(n - k, k), \quad n \geq 2 \text{ and } 1 \leq k \leq n.$$

The first two lines are obvious. As for the recurrence relation in the third line, there are two possibilities for the least part, λ_k, in a partition of n into k parts: either $\lambda_k = 1$ or $\lambda_k > 1$. If $\lambda_k = 1$, then there are $p(n - 1, k - 1)$ partitions of the remaining number $n - 1$ into $k - 1$ parts. If $\lambda_k > 1$, then partitions of n into k parts are in one-to-one correspondence with partitions of $n - k$ into k parts (subtract 1 from each part in each partition of n).

These recurrence relations and the formula $p(n) = \sum_{k=1}^{n} p(n, k)$ yield Tables 15.1 and 15.2. We should note that in order to calculate $p(n)$ using this method, we must calculate an entire table of values of $p(n, k)$. A more direct way to calculate $p(n)$ for large n is furnished by Euler's pentagonal number theorem, which we will see later.

We can picture partitions with "Ferrers diagrams," named after Norman Ferrers (1829–1903). Given a partition $n = \lambda_1 + \lambda_2 + \cdots + \lambda_k$, with $\lambda_1 \geq \lambda_2 \geq \cdots \geq \lambda_k$, the corresponding Ferrers diagram consists of k rows of dots with λ_i dots in row i, for $1 \leq i \leq k$. The Ferrers diagram for the partition $11 = 7 + 2 + 1 + 1$ is shown in Figure 15.1.

$$
\begin{array}{ll}
\lambda_1 & \bullet \; \bullet \; \bullet \; \bullet \; \bullet \; \bullet \; \bullet \\
\lambda_2 & \bullet \; \bullet \\
\lambda_3 & \bullet \\
\lambda_4 & \bullet
\end{array}
$$

FIGURE 15.1: The Ferrers diagram of a partition of 11.

We create the *transpose* of a Ferrers diagram by writing each row of dots

n \ k	1	2	3	4	5	6	7	8	9	10
1	1									
2	1	1								
3	1	1	1							
4	1	2	1	1						
5	1	2	2	1	1					
6	1	3	3	2	1	1				
7	1	3	4	3	2	1	1			
8	1	4	5	5	3	2	1	1		
9	1	4	7	6	5	3	2	1	1	
10	1	5	8	9	7	5	3	2	1	1

TABLE 15.1: Partition numbers $p(n, k)$.

as a column. We call the resulting partition the *conjugate* of the original partition. For example, the partition $11 = 7 + 2 + 1 + 1$ of Figure 15.1 is transposed to create the conjugate partition $11 = 4 + 2 + 1 + 1 + 1 + 1 + 1$ of Figure 15.2.

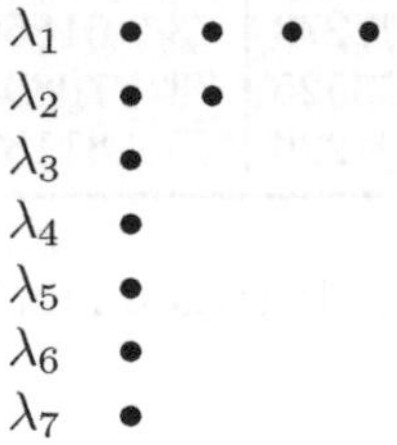

FIGURE 15.2: A transpose Ferrers diagram.

You may want to test your understanding by matching each partition of 5 in Example 15.1 with its conjugate. How many of the partitions are self-conjugate?

Let $p(0) = 1$. Then the (ordinary) generating function for the partition numbers $p(n)$ is

$$\sum_{n=0}^{\infty} p(n)x^n = 1 + x + 2x^2 + 3x^3 + 5x^4 + 7x^5 + 11x^6 + 15x^7 + 22x^8 + \cdots.$$

This generating function has a representation as an infinite product.

n	$p(n)$	n	$p(n)$	n	$p(n)$	n	$p(n)$
1	1	26	2436	51	239943	76	9289091
2	2	27	3010	52	281589	77	10619863
3	3	28	3718	53	329931	78	12132164
4	5	29	4565	54	386155	79	13848650
5	7	30	5604	55	451276	80	15796476
6	11	31	6842	56	526823	81	18004327
7	15	32	8349	57	614154	82	20506255
8	22	33	10143	58	715220	83	23338469
9	30	34	12310	59	831820	84	26543660
10	42	35	14883	60	966467	85	30167357
11	56	36	17977	61	1121505	86	34262962
12	77	37	21637	62	1300156	87	38887673
13	101	38	26015	63	1505499	88	44108109
14	135	39	31185	64	1741630	89	49995925
15	176	40	37338	65	2012558	90	56634173
16	231	41	44583	66	2323520	91	64112359
17	297	42	53174	67	2679689	92	72533807
18	385	43	63261	68	3087735	93	82010177
19	490	44	75175	69	3554345	94	92669720
20	627	45	89134	70	4087968	95	104651419
21	792	46	105558	71	4697205	96	118114304
22	1002	47	124754	72	5392783	97	133230930
23	1255	48	147273	73	6185689	98	150198136
24	1575	49	173525	74	7089500	99	169229875
25	1958	50	204226	75	8118264	100	190569292

TABLE 15.2: Partition numbers $p(n)$.

Theorem 15.2.

$$\sum_{n=0}^{\infty} p(n)x^n = \prod_{k=1}^{\infty} (1 - x^k)^{-1}$$

The concept behind this formula is basically the same as that of the generating functions for making change and for integer triangles that we saw in Chapters 7 and 8.

Denote by $p(n \mid \lambda_1 = k)$ the number of partitions of n in which the largest part is k. The next proposition is a simple observation using transpose Ferrers diagrams.

Proposition 15.3.

$$p(n, k) = p(n \mid \lambda_1 = k)$$

Denote by $p(n, \leq k)$ the number of partitions of n into at most k parts, and by $p(n \mid \lambda_1 \leq k)$ the number of partitions of n into parts of size at most k. From Proposition 15.3, we have

$$p(n, \leq k) = p(n \mid \lambda_1 \leq k).$$

Each partition of n into k parts corresponds to a partition of $n - k$ into at most k parts (by subtracting 1 from each part). From the previous relation, we conclude that

$$p(n, k) = p(n - k \mid \lambda_1 \leq k).$$

That is to say, the number of partitions of n into k parts is equal to the number of partitions of $n - k$ with largest part at most k. The generating function for the partition numbers $p(n, k)$ follows immediately.

Theorem 15.4.

$$\sum_{n=k}^{\infty} p(n, k)x^n = x^k \prod_{j=1}^{k} (1 - x^j)^{-1}$$

We have covered the basic identities and generating functions for partitions. Now let's consider a variety of restrictions on partitions. Denote by $p(n \mid \text{distinct parts})$ the number of partitions of n into distinct parts. The next result is a simple observation regarding another infinite product representation.

Proposition 15.5.

$$\sum_{n=0}^{\infty} p(n \mid \text{distinct parts})x^n = \prod_{k=1}^{\infty} (1 + x^k)$$

Denote by $p(n \mid \text{odd parts})$ the number of partitions of n into parts each of which is an odd number. Leonhard Euler (1707–1783) proved the following identity as one of many about partitions.

Proposition 15.6 (L. Euler, 1748).

$$p(n \mid \text{odd parts}) = p(n \mid \text{distinct parts})$$

Proof. The proof is a bit of generating function magic. We have

$$\sum_{n=0}^{\infty} p(n \mid \text{odd parts})x^n = \frac{1}{(1-x)(1-x^3)(1-x^5)\ldots}$$

$$= \frac{(1-x^2)}{(1-x)(1-x^2)} \cdot \frac{(1-x^4)}{(1-x^3)(1-x^4)} \cdot \frac{(1-x^6)}{(1-x^5)(1-x^6)} \cdots$$

$$= \frac{(1-x^2)}{(1-x)} \cdot \frac{(1-x^4)}{(1-x^2)} \cdot \frac{(1-x^6)}{(1-x^3)} \cdots$$

$$= (1+x)(1+x^2)(1+x^3)\ldots$$

$$= \sum_{n=0}^{\infty} p(n \mid \text{distinct parts})x^n.$$

The result follows by comparing coefficients of x^n. ∎

Example 15.7. Find the partitions of 6 into odd parts and the partitions of 6 into distinct parts.

Solution:

$p(6 \mid \text{odd parts}) = 4 \qquad (5+1, \ 3+3, \ 3+1+1+1, \ 1+1+1+1+1+1)$

$p(6 \mid \text{distinct parts}) = 4 \qquad (6, \ 5+1, \ 4+2, \ 3+2+1)$

∎

Denote by $p(n \mid \text{even } \# \text{ distinct parts})$ the number of partitions of n into an even number of distinct parts, and by $p(n \mid \text{odd } \# \text{ distinct parts})$ the number of partitions of n into an odd number of distinct parts. The following result is similar to the kind of generating function that we found in Chapter 7 for the number of ways of making change with an odd number or an even number of coins.

Proposition 15.8.

$$\prod_{n=1}^{\infty} (1-x^n) = \sum_{n=0}^{\infty} [p(n \mid \text{even } \# \text{ distinct parts}) - p(n \mid \text{odd } \# \text{ distinct parts})]x^n$$

Each coefficient of x^n in the expansion of the above product is 0, 1, or -1. The pattern is given succinctly in terms of "pentagonal numbers." A *pentagonal number* is a number that can be represented by dots arranged in a pentagonal array, as in Figure 15.3. Thus, the pentagonal numbers are

$$1, \ 5, \ 12, \ 22, \ 35, \ 51, \ 70, \ 92, \ 117, \ \ldots.$$

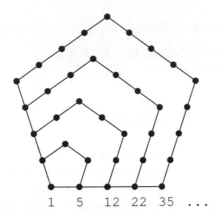

1 5 12 22 35 ...

FIGURE 15.3: Pentagonal numbers.

Let a_k be the kth pentagonal number. We observe a recurrence formula for $\{a_k\}$ from the diagram:

$$a_k = a_{k-1} + 3k - 2, \quad k \geq 2, \quad a_1 = 1.$$

Hence

$$a_k = a_1 + 3(2 + 3 + \cdots + k) - 2(k - 1)$$

$$= 1 + 3\left(\frac{k(k+1)}{2} - 1\right) - 2(k - 1)$$

$$= \frac{k(3k - 1)}{2},$$

and we see that a pentagonal number has the form

$$k(3k - 1)/2, \quad k \geq 1.$$

We may replace k by $-k$, obtaining the so-called "pseudopentagonal numbers":

$$(-k)(-3k - 1)/2 = k(3k + 1)/2, \quad k \geq 1.$$

Thus, the pseudopentagonal numbers are

$$2, \ 7, \ 15, \ 26, \ 40, \ 57, \ 77, \ 100, \ 126, \ \ldots.$$

The two types of pentagonal numbers, together with 0, are called "generalized pentagonal numbers" and are given by

$$n = k(3k \pm 1)/2, \quad k \geq 0.$$

Proposition 15.9.

$p(n \mid \text{even number of distinct parts}) - p(n \mid \text{odd number of distinct parts})$

$$= \begin{cases} (-1)^k & \text{if } n = k(3k \pm 1)/2, \quad \text{for } k \geq 1 \\ 0 & \text{otherwise} \end{cases}$$

The following bijective proof using Ferrers diagrams was discovered in 1881 by Fabian Franklin (1853–1939).

Proof. We will describe a correspondence between partitions of n with an even number of distinct parts and partitions of n with an odd number of distinct parts. The correspondence is a one-to-one correspondence for all n that are not generalized pentagonal numbers. For generalized pentagonal numbers, there is one extra partition in one of the collections.

Given a Ferrers diagram for a partition with distinct parts, let H (for horizontal) be the bottom row of dots (representing the smallest part of the partition), and let D (for diagonal) be the longest diagonal of dots, starting with the right-most dot in the top row. Let h be the number of dots in H and d the number of dots in D. If $h \leq d$, then move H so that it forms a diagonal to the right of D, with its top dot in the top row. If $h > d$, then move D to the bottom row as a new smallest part.

Figure 15.4 illustrates the case $n = 17$ (not a generalized pentagonal number), showing the correspondence between the Ferrers diagrams of the partitions $17 = 6+5+4+2$ (even number of distinct parts) and $17 = 7+6+4$ (odd number of distinct parts).

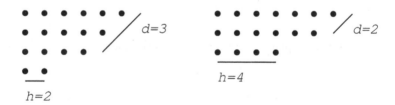

FIGURE 15.4: Correspondence between partitions with distinct parts.

Usually (as in our example), the correspondence changes the parity of the number of parts in the partition, giving a one-to-one correspondence between partitions with an even number of distinct parts and partitions with an odd number of distinct parts. The correspondence fails in two cases: when H and D have a dot in common and $h = d$ or $h = d + 1$. In the first case,

$$n = d^2 + 1 + \cdots + (d-1) = \frac{d(3d-1)}{2}.$$

In the second case,

$$n = d^2 + 1 + \cdots + d = \frac{d(3d+1)}{2}.$$

We see that these two cases can occur only when n is a generalized pentagonal number. If n is a dth generalized pentagonal number, then the difference between the number of partitions with an even number of distinct parts and the number of partitions with an odd number of distinct parts is $(-1)^d$. ∎

Example 15.10. Verify Proposition 15.9 for $n = 12$. Note that $12 = 3 \cdot (3 \cdot 3 - 1)/2$ is a pentagonal number.

Solution:

$p(12 \mid \text{even number of distinct parts}) = 7$

$(11 + 1,\ 10 + 2,\ 9 + 3,\ 8 + 4,\ 7 + 5,\ 6 + 3 + 2 + 1,\ 5 + 4 + 2 + 1)$

$p(12 \mid \text{odd number of distinct parts}) = 8$

$(12,\ 9 + 2 + 1,\ 8 + 3 + 1,\ 7 + 4 + 1,\ 6 + 5 + 1,\ 7 + 3 + 2,\ 6 + 4 + 2,\ 5 + 4 + 3)$

$p(12 \mid \text{even number of distinct parts}) - p(12 \mid \text{odd number of distinct parts})$

$= 7 - 8 = (-1)^3$

∎

Combining Propositions 15.8 and 15.9, we obtain Euler's famous pentagonal number theorem.

Theorem 15.11 (Euler's Pentagonal Number Theorem, 1750).

$$\prod_{n=1}^{\infty}(1 - x^n) = \sum_{k=-\infty}^{\infty} (-1)^k x^{k(3k-1)/2}$$

From the relation

$$1 = \sum_{n=0}^{\infty} p(n)x^n \prod_{n=1}^{\infty}(1 - x^n),$$

we obtain a fast recurrence formula for partition numbers.

Corollary 15.12.

$p(0) = 1$

$p(n) = p(n-1) + p(n-2) - p(n-5) - p(n-7) + p(n-12) + p(n-15) - \cdots$

The formula in the recurrence relation consists of all partition numbers $p(n-m)$, where m is a generalized pentagonal number such that $n-m \geq 0$. These terms occur in pairs with the same sign, one pentagonal and the other pseudopentagonal. The signs of the pairs alternate, with the positive signs occurring when k is odd and the negative signs when k is even, where m is a kth generalized pentagonal number.

Example 15.13. Find $p(10)$ using the above recurrence formula.

Solution: We have

$$p(1) = \mathbf{1}$$
$$p(2) = p(1) + p(0) = 1 + 1 = \mathbf{2}$$
$$p(3) = p(2) + p(1) = 2 + 1 = \mathbf{3}$$
$$p(4) = p(3) + p(2) = 3 + 2 = \mathbf{5}$$
$$p(5) = p(4) + p(3) - p(0) = 5 + 3 - 1 = \mathbf{7}$$
$$p(6) = p(5) + p(4) - p(1) = 7 + 5 - 1 = \mathbf{11}$$
$$p(7) = p(6) + p(5) - p(2) - p(0) = 11 + 7 - 2 - 1 = \mathbf{15}$$
$$p(8) = p(7) + p(6) - p(3) - p(1) = 15 + 11 - 3 - 1 = \mathbf{22}$$
$$p(9) = p(8) + p(7) - p(4) - p(2) = 22 + 15 - 5 - 2 = \mathbf{30}$$
$$p(10) = p(9) + p(8) - p(5) - p(3) = 30 + 22 - 7 - 3 = \mathbf{42}.$$

These values agree with Table 15.2. ■

With a computer, we can use the recurrence relation from Euler's pentagonal number theorem to compute $p(n)$ for large n. For example,

$$p(1000) = 24061467864032622473692149727991 \doteq 2.4 \times 10^{31}.$$

The asymptotic formula of Hardy and Ramanujan gives an estimate approximately 1.014 times this number.

Exercises

1. List the partitions of 5, 6, and 7.

◇2. Implement recurrence formulas to calculate $p(n, k)$ and $p(n)$.

⋆3. Find formulas for $p(n, 1)$, $p(n, 2)$, and $p(n, 3)$. Conjecture an asymptotic estimate for $p(n, k)$ with k fixed.

◇4. Use a computer to verify Theorem 15.2 for $0 \leq n \leq 20$.

†5. Prove Theorem 15.2.

6. Prove Proposition 15.5.

7. Prove Proposition 15.6 by describing a one-to-one correspondence between partitions of n into odd parts and partitions of n into distinct parts.

8. Denote by $p(n \mid \text{even} \# \text{ of parts})$ and $p(n \mid \text{odd} \# \text{ of parts})$ the number of partitions of n into an even number of parts and into an odd number of parts, respectively. Denote by $p(n \mid \text{distinct odd parts})$ the number of partitions of n with distinct odd parts. Let $\tilde{p}(n)$ be the number of self-conjugate partitions of n. Prove that

$$\tilde{p}(n) = p(n \mid \text{distinct odd parts})$$

$$= (-1)^n (p(n \mid \text{even} \# \text{ of parts}) - p(n \mid \text{odd} \# \text{ of parts})).$$

9. Show that the number of partitions of n in which no part occurs exactly once is the same as the number of partitions of n in which none of the parts is congruent to 1 or 5 modulo 6.

10. Prove that the two sequences of generalized pentagonal numbers have no elements in common.

11. Prove that every pentagonal number is $1/3$ of a triangular number, i.e., a number of the form $1 + 2 + 3 + \cdots + n = n(n+1)/2$.

12. Show the correspondence in Proposition 15.9 for the partitions with distinct parts of $n = 10$ (not a generalized pentagonal number).

◇13. Use the recurrence relation from Euler's pentagonal number theorem to compute $p(1000)$.

Part VI

Information Theory

Chapter 16

What Is Surprise?

The memory feat of reciting the order of a shuffled deck of 52 cards is worth about 223 bits.

The subject of information theory, of which entropy is the central concept, was born in 1948 when Claude E. Shannon (1914–2001) published his landmark paper [Sha48] on information sources and channels. The field has since borne many beautiful results both pure and applied. Information theory is the practical motivation for error-correcting codes.

To introduce this topic, let's begin by considering some memory feats. Suppose that you take a deck of 52 cards, shuffle it, turn each card face up briefly, and then turn it face down again. Suppose that after you have seen every card you can recite the entire sequence of cards in order. (This memory stunt is within the range of most people, with practice.) Now, suppose that your friend has a different memory feat. She can memorize a string of 100 digits that you read at random from, say, a telephone book. After she has heard the digits, she can recite them back in order. These are two impressive memory feats, but which one is better? Can we put a measure on these stunts, or are we just trying to equate apples and oranges?

It turns out that we can measure these memory feats and the measurement is the key ingredient for the concept of mathematical surprise. There are 52! different ways to order a deck of 52 cards. Let's denote the "value" of memorizing a deck of cards in order as 52!. This number is gigantic:

$$52! = 80658175170943878571660636856403766$$

$$975289505440883277824000000000000.$$

In fact, having 68 digits, this number is between 10^{67} and 10^{68}. Regarding the other memory stunt, memorizing 100 digits, there are 10^{100} different strings of 100 digits, so let's say that the value of this feat is 10^{100}. Thus, the digit-memorizing feat is far better than the card-memorizing feat.

Our measure for the memory feats is simply the number of possible outcomes of the events in question. If S is the set of possible outcomes, and we are able to produce (e.g., memorize) one of them, then we have

123

given the value of $|S|$ to that feat. Similar calculations hold for probability spaces. If the sample space (set of possible outcomes) of an event is S, and all simple events are equally likely, then the probability that a particular outcome occurs is $p = 1/|S|$. According to our definition, the value assigned to this outcome is $1/p$.

Various synonyms describe the quantity that we are defining. This quantity is referred to as "surprise," "information," and "uncertainty." Typically, we normalize our quantity by taking a logarithm base 2. Suppose that E is an event that occurs with probability $p > 0$. If we learn with certainty that E has occurred, then we say that we have received

$$I(p) = \log_2(1/p) = -\log_2 p$$

bits of *information*. The reason for the base 2 is that this is the number of bits it takes to represent the value in binary.

Example 16.1. What is the amount of surprise, in bits, of the memory stunt of reciting the order of a shuffled deck of 52 cards?

Solution: The value of the stunt is $\log_2(52!) \doteq 223$ bits.　　■

We justify the definition of surprise by observing that I, a function from $(0, 1]$ into $\mathbf{R}^+ \cup \{0\}$, has three properties that can be interpreted as common-sense statements about information.

(1) $I(p) \geq 0$ for all $p \in (0, 1]$.

Interpretation: Suppose that E is an event that occurs with probability p. If we learn that E has occurred, then we certainly have not lost information.

(2) $I(p)$ is a continuous function of p.

Interpretation: If the likelihood of E varies slightly, then the information associated with E varies only slightly.

(3) $I(pq) = I(p) + I(q)$ for all p, q.

Interpretation: Suppose that E and F are independent events with $\Pr(E) = p$, $\Pr(F) = q$, and $\Pr(EF) = pq$. If we already know that E has occurred and we are told that F occurs, then the new information obtained is $I(q) = I(pq) - I(p)$.

Theorem 16.2. If I is a function from $(0, 1]$ into $\mathbf{R}$ satisfying the properties (1), (2), and (3) above, then $I(p) = -C \log_2 p$, where C is an arbitrary positive number.

Proof. Let $p \in (0, 1]$. From the property (3), $I(p^2) = I(p) + I(p) = 2I(p)$, and, by mathematical induction, $I(p^m) = mI(p)$ for all positive integers m. Similarly, $I(p) = I(p^{1/n} \cdot \cdots \cdot p^{1/n}) = nI(p^{1/n})$, and hence $I(p^{1/n}) = (1/n)I(p)$ for all integers n. These observations imply that $I(p^{m/n}) = (m/n)I(p)$. Hence, by (2), $I(p^x) = xI(p)$ for all positive real numbers x. Therefore $I(p) = I((1/2)^{-\log_2 p}) = -I(1/2)\log_2 p = -C\log_2 p$, where $C = I(1/2)$. By (1), C must be positive.　　■

For convenience, we take $C = 1$, so that $I(p) = -\log_2 p$.

Here are two immediate consequences of the formula for $I(p)$:

(1) $I(1)=0$.

Interpretation: If an event E occurs with probability 1 and we are told that E occurs, then we have gained no information.

(2) $I(p)$ is a strictly decreasing function of p; that is, $p < q$ implies that $I(p) > I(q)$.

Interpretation: If an event E is less likely than an event F, then we receive more information if we are told that E occurs than if we are told that F occurs.

Shannon's seminal idea in information theory is to associate an amount of information, or entropy, with an information source. Now that we have defined a function that measures the uncertainty (also called surprise or information) of an event, we will proceed to show how to define the entropy of an information source.

A *source* S is a sequence of random variables X_1, X_2, X_3, ... with common range $\{x_1, \ldots, x_n\}$. Such a sequence is also called a *discrete-time, discrete-valued stochastic process*. ("Stochastic" comes from the Greek word *stokhastikos*, meaning "capable of aiming, conjectural," and here we are forming conjectures about the source from values of some of its terms.) The elements x_i are called *states* or *symbols*, and we think of S as emitting these symbols at regular intervals of time, as in Figure 16.1.

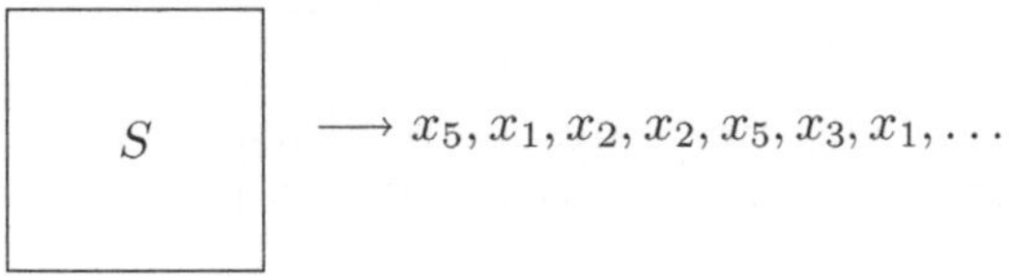

$$\boxed{ S } \longrightarrow x_5, x_1, x_2, x_2, x_5, x_3, x_1, \ldots$$

FIGURE 16.1: A source.

In a *discrete memoryless source* (DMS), also called a *zero-memory* source, the random variables X_i are independent and identically distributed. Suppose that the X_i are distributed as X, a random variable that takes values x_1, ..., x_n with probabilities p_1, ..., p_n, respectively, where each $p_i \geq 0$ and $\sum_{i=1}^{n} p_i = 1$. As a random variable, X is a function from $\mathbf{N}$ into $\{x_1, \ldots, x_n\}$, and $\Pr\{X(i) = x_j\} = p_j$, for all $i \geq 1$ and $1 \leq j \leq n$. The sequence X_1, X_2, X_3, ..., called a *sampling* of X, represents a process that repeatedly selects values of X. If S is a discrete memoryless source, we write

$$S = \begin{pmatrix} x_1 & x_2 & \cdots & x_n \\ p_1 & p_2 & \cdots & p_n \end{pmatrix}.$$

Given our formula for I, we can say that when S emits the symbol x_i, we receive $I(p_i) = -\log_2 p_i$ bits of information. Since x_i is emitted with probability p_i, the average amount of information obtained per symbol is

$$H(S) = \sum_{i=1}^{n} p_i I(p_i) = -\sum_{i=1}^{n} p_i \log_2 p_i \text{ bits.}$$

We call $H(S)$ the *entropy* of the source S. The concept of entropy was introduced by Ludwig Boltzmann (1844–1906) in 1896, but Shannon was the first to apply it to information sources.

We will usually not mention the units ("bits") when referring to information or entropy. We will sometimes suppress the index in the sigma notation. Also, we write log for $\log_2$ and set $0 \log 0 = \lim_{x \to 0+} x \log x = 0$.

Note that H is a continuous function of the p_i. Some other properties of H will be demonstrated in the following discussion.

Example 16.3. Suppose that a source S is given by

$$S = \begin{pmatrix} x_1 & x_2 & x_3 & x_4 \\ 1/2 & 1/4 & 1/8 & 1/8 \end{pmatrix}.$$

What is the entropy of S?

Solution: The entropy of S is

$$H(S) = -\frac{1}{2}\log\frac{1}{2} - \frac{1}{4}\log\frac{1}{4} - \frac{1}{8}\log\frac{1}{8} - \frac{1}{8}\log\frac{1}{8} = \frac{1}{2} + \frac{2}{4} + \frac{3}{8} + \frac{3}{8} = \frac{7}{4}.$$

∎

Sources consisting of only two states are particularly important. Let S be the source with states x_1 and x_2, occurring with probabilities p and q, respectively:

$$S = \begin{pmatrix} x_1 & x_2 \\ p & q \end{pmatrix}.$$

The entropy of S is $H(S) = -p \log p - q \log q$. Deliberately over-using the symbol H, we denote this expression by $H(p)$. Figure 16.2 shows the graph of this function.

Here is a technical lemma that will be used three times in the upcoming proofs of facts about entropy.

Lemma 16.4 (Convexity of the Logarithm Function). Let $p_1, \ldots, p_n$ and $q_1, \ldots, q_n$ be nonnegative real numbers with $\sum_{i=1}^{n} p_i = \sum_{i=1}^{n} q_i = 1$. Then

$$-\sum_{i=1}^{n} p_i \log p_i \leq -\sum_{i=1}^{n} p_i \log q_i,$$

with equality if and only if $p_i = q_i$ for all i.

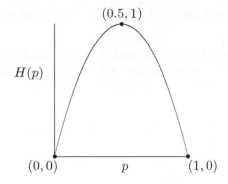

FIGURE 16.2: The entropy function $H(p)$.

Proof. There is no contribution to the summations from any $p_i = 0$, so we discard these values. If any of the remaining q_i are 0, then the inequality holds, so we may assume that no q_i is 0. Now we obtain

$$
\begin{aligned}
\sum p_i \log q_i - \sum p_i \log p_i &= \sum p_i(\log q_i - \log p_i) \\
&= \sum p_i \log(q_i/p_i) \\
&\leq \sum p_i(q_i/p_i - 1) \\
&= \sum q_i - \sum p_i \\
&= \sum q_i - 1 \\
&\leq 0.
\end{aligned}
$$

The first inequality follows from the fact that $\log x \leq x - 1$ for $x \in \mathbf{R}^+$. Equality occurs if and only if $x = 1$, i.e., when $p_i = q_i$ for all i. ∎

We next consider two extreme types of sources.

A source is *uniform* if every state has the same likelihood of occurring. The uniform source with n states is

$$
S = \left(\begin{array}{ccccc} x_1 & \cdots & x_i & \cdots & x_n \\ 1/n & \cdots & 1/n & \cdots & 1/n \end{array} \right).
$$

It has entropy $-\sum_{i=1}^{n}(1/n) \log(1/n) = \log n$.

A source is *singular* if $p_i = 1$ for some i. The singular source with n states is

$$
S = \left(\begin{array}{cccccccc} x_1 & \cdots & x_{i-1} & x_i & x_{i+1} & \cdots & x_n \\ 0 & \cdots & 0 & 1 & 0 & \cdots & 0 \end{array} \right).
$$

It has entropy 0, since $0 \log 0 = 1 \log 1 = 0$. A singular source conveys no information.

Theorem 16.5. Let S be a source with n states. Then $0 \le H(S) \le \log n$, with the lower bound attained if and only if S is singular and the upper bound attained if and only if S is uniform.

Proof. To establish the lower bound, we note that $p_i \log p_i > 0$ if $p_i \in (0, 1)$. Therefore $H(S) > 0$ unless each $p_i = 0$ or 1, in which case S is singular and $H(S) = 0$, as stated above.

We now prove the upper bound.

$$
\begin{aligned}
H(S) &= -\sum p_i \log p_i \\
&\le -\sum p_i \log \frac{1}{n} \quad \text{(by Lemma 16.4)} \\
&= -\log \frac{1}{n} \sum p_i \\
&= \log n.
\end{aligned}
$$

According to Lemma 16.4, equality occurs if and only if $p_i = 1/n$ for all i, i.e., the source is uniform. ∎

Although the remaining discussion will focus on discrete memoryless sources, let us say a few words about a more sophisticated source called a *Markov source*.

One everyday example of a source is a person speaking or writing English. The DMS model that we have discussed allows us to make a simplistic calculation of the entropy of the English language. Based on a memoryless source of 27 characters (26 letters and a space), the DMS model yields an entropy value of about 4 (each revealed character conveys 4 bits of information). But the model is too simplistic. If past characters are used to predict future ones (the source has a memory), then the calculated entropy of English decreases to a number believed to be between 0.5 and 1.5.

One model that takes into account past symbols is called a Markov source. For realistic computations, we use an mth order Markov source, i.e., a source in which the probability that a given symbol is emitted depends on the previous m symbols. Formally, an mth order *Markov source* (or *Markov chain*) consists of an alphabet $A = \{x_1, \dots, x_n\}$ and the values of conditional probabilities (also called *transitional probabilities*)

$$
\Pr(x_i : x_{j_1}, \dots, x_{j_m}),
$$

for $i = 1, \dots, n$ and $(j_1, \dots, j_m) \in [n]^m$. We say that m consecutive symbols of an mth order Markov source constitute a *state* of the source. Since

S contains n symbols, the source has n^m states, and these may be represented in a state diagram as in the next example.

Example 16.6. Let S be a first-order Markov source with alphabet $A = \{a, b\}$ and conditional probabilities $\Pr(a : a) = 0.6$, $\Pr(b : b) = 0.6$, $\Pr(b : a) = 0.4$, and $\Pr(a : b) = 0.4$. Show the state diagram of this source.

Solution: The state diagram of the source is shown in Figure 16.3.

FIGURE 16.3: A first-order Markov source.

∎

An mth order Markov source can always be encoded as a first-order Markov source (take the symbols of the new source to be m-tuples of the old symbols). Let us calculate the entropy of a first-order Markov source. If the source is in state x_i, then the information obtained when the symbol x_j occurs is

$$I(x_j : x_i) = -\log \Pr(x_j : x_i).$$

Therefore the average amount of information conveyed when a state is revealed after x_i is

$$H(S : x_i) = -\sum_A \Pr(x_j : x_i) \log \Pr(x_j : x_i).$$

This means that the entropy of the source is

$$H(S) = -\sum_{A^2} \Pr(x_i) \Pr(x_j : x_i) \log \Pr(x_j : x_i).$$

Example 16.7. What is the entropy of the source S in the previous example?

Solution: First, we find the steady-state probabilities $\Pr(a)$ and $\Pr(b)$. Although we could solve the system of equations

$$\Pr(a) + \Pr(b) = 1$$

$$\Pr(a) = 0.6\Pr(a) + 0.4\Pr(b)$$

$$\Pr(b) = 0.4\Pr(a) + 0.6\Pr(b),$$

it is easier to note that, by symmetry, $\Pr(a) = \Pr(b) = 0.5$. Next, we calculate

$$
\begin{aligned}
H(S) \;=\; & -0.5 \cdot (0.6\log 0.6) - 0.5 \cdot (0.4\log 0.4) \\
& -0.5 \cdot (0.6\log 0.6) - 0.5 \cdot (0.4\log 0.4) \\
\doteq \;& 0.970951.
\end{aligned}
$$

Notice that the entropy of this source is less than 1. By comparison, a memoryless source (i.e., a zero-order Markov source) that takes values a and b with equal probabilities has entropy equal to 1. The first-order Markov source has lower entropy than the memoryless source because knowledge of previous symbols reduces uncertainty. To put it into a common-place setting, let a stand for "fair day" and b stand for "foul day." Then the Markov source is a very simplistic model of weather, in which current weather is somewhat influenced by the previous day's weather. Although in the long run the model says that the weather is fair half of the days and foul half of the days, the uncertainty in the weather is less than it would be if the current weather had no relation to the previous day's weather. ■

Exercises

1. A memory performer memorizes and recites the order of two decks of cards shuffled together. What is the information content of this stunt?

◇2. Another memory performer can recite a string of 100 arbitrary letters. Does this feat, or the one in the previous exercise, have a greater value?

3. Professor Bumble says that if told a digit (0 through 9) and a letter (A through Z), he can recite them back, but not always in the order given. What is his stunt worth in bits?

4. If an event occurs with probability 2^{-10}, and we learn that the event occurs, how many bits of information do we receive? If we learn that the event does not occur, how many bits of information do we receive?

5. Two dice are rolled and you are told that their total is 9. If you are then told that both dice show an even number, how much additional information is this?

6. Prove that $I(pq) = I(p) + I(q)$, for all $0 \le p, q \le 1$.

7. Consider a source with states a, b, c, d, e, f, with probabilities $1/2$, $1/4$, $1/8$, $1/16$, $1/32$, and $1/32$ (respectively). Find the entropy.

8. Consider a source with states a, b, c, d, with probabilities $1/3$, $1/3$, $1/6$, $1/6$ (respectively). Find the entropy.

9. Give an example of a source with five states and entropy $15/8$ bits.

10. What is the minimum number of states in a source with entropy 10 bits?

11. Let S be a source with states x_1, x_2, $\ldots$, x_n, with probabilities p_1, p_2, $\ldots$, p_n (respectively). Let T be a source with states x_1, x_2, $\ldots$, x_n, x_{n+1}, with probabilities $\frac{1}{2}p_1$, $\frac{1}{2}p_2$, $\ldots$, $\frac{1}{2}p_n$, $\frac{1}{2}$ (respectively). Find a formula for the entropy of T in terms of the entropy of S.

12. Suppose that

$$S = \left(\begin{array}{cc} x_1 & x_2 \\ \alpha & \beta \end{array} \right),$$

with $\alpha + \beta = 1$ and α, $\beta \ge 0$. Use calculus to show that $H(S)$ is maximized when $\alpha = \beta = 0.5$.

13. Professor Bumble has a class in which he sometimes gives pop quizzes. If he gives a pop quiz one day, there is a $2/5$ chance that he will give a pop quiz the next day. If he does not give a pop quiz one day, there is a $2/5$ chance that he will not give a pop quiz the next day. At the beginning of class, a student remarks that he knows that there will be a pop quiz today. If this is true, how much information is it?

14. Let S be a first-order Markov source with alphabet $A = \{a, b, c\}$ and conditional probabilities $\Pr(a : a) = \Pr(b : b) = \Pr(c : c) = 0.1$, $\Pr(b : a) = \Pr(c : b) = \Pr(a : c) = 0.7$, and $\Pr(a : b) = \Pr(b : c) = \Pr(c : a) = 0.2$. Find $H(S)$.

15. If the Markov source of the previous exercise is changed so that $\Pr(b : a) = \Pr(c : b) = \Pr(a : c) = \Pr(a : b) = \Pr(b : c) = \Pr(c : a) = 0.45$, will the entropy increase or decrease? Give a common sense argument rather than a calculation.

16. Let S be a first-order Markov source with alphabet $A = \{a, b\}$ and conditional probabilities $\Pr(a : a) = \Pr(b : b) = 0.1$ and $\Pr(b : a) = \Pr(a : b) = 0.9$. Let T be a first-order Markov source with alphabet $A = \{a, b\}$ and conditional probabilities $\Pr(a : a) = \Pr(b : b) = 0.9$ and $\Pr(b : a) = \Pr(a : b) = 0.1$. Let U be a first-order Markov source with alphabet $A = \{a, b\}$ and conditional probabilities $\Pr(a : a) = \Pr(a : b) = 0.9$ and $\Pr(b : b) = \Pr(b : a) = 0.1$. Find the entropies of S, T, and U.

17. Select a page of text in English and calculate its entropy as a memoryless source. Assume that there are only 26 possible symbols, the letters of the alphabet.

Chapter 17

A Coin-Tossing Game

We play a game in which we are given a biased coin with probability 0.53 of landing heads and probability 0.47 of landing tails. We start the game with $1 and once a day, every day, we may bet any fraction of our current amount on the event that the coin lands heads. If the coin lands heads, we win an even money payoff; if the coin lands tails, we lose the amount bet. If we lose all of our money, the game is over. What fraction of our current amount should we bet every day to maximize our long-term profit?

Let's solve the above problem in general, where the coin lands heads with probability p and tails with probability q, with $p+q = 1$ and $p > q > 0$. The following analysis was demonstrated in 1956 by John L. Kelly Jr. (1923–1965) [Kel56].

We shouldn't wager all of our current amount at any time, because we could lose and not be able to continue playing. Of course, it doesn't make sense to bet on the coin landing tails (even if we could) because the coin is biased toward heads. However, let's imagine the possibility of placing bets on heads *and* tails.

If we divide our entire fortune on the two types of bets, some of the bets will cancel out. A hunch is that we should bet p of our current amount on heads and q of our current amount on tails. In this case, q of the bets will cancel out, leaving a remainder of $p - q$ of our amount bet on heads. This is equivalent to betting $p - q$ of our amount on heads and nothing on tails. With $p = 0.53$, we would bet $0.53 - 0.47 = 0.06$ of our current amount on heads.

Let's prove that the hunch results in the maximum long-term profit. Over n days, the expected number of heads is pn and the expected number of tails is qn. Suppose that we bet λ_1 of our amount on heads and λ_2 of our amount on tails, where $\lambda_1 + \lambda_2 = 1$. We wish to show that the best choice is $\lambda_1 = p$ and $\lambda_2 = q$. Each occurrence of heads yields a return of $2\lambda_1$ of our amount, while each occurrence of tails yields a return of $2\lambda_2$ of our amount. Hence, after n days, our expected amount is

$$(2\lambda_1)^{np}(2\lambda_2)^{nq}.$$

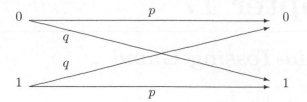

FIGURE 17.1: The Binary Symmetric Channel (BSC).

We write this expression as an exponential function,

$$2^{cn},$$

where
$$c = 1 + p \log \lambda_1 + q \log \lambda_2,$$

and logarithms are base 2.

The hunch is that c, the *coefficient of growth*, is maximized when $\lambda_1 = p$ and $\lambda_2 = q$. From Lemma 16.4 (convexity of the log function),

$$c \leq 1 + p \log p + q \log q,$$

with equality if and only if $\lambda_1 = p$ and $\lambda_2 = q$. Therefore, the optimal strategy is to consistently bet $p - q$ of our current amount.

With $p = 0.53$, the maximum coefficient of growth is $c \doteq 0.00259841$. At this growth rate, it would take about 21 years of steadily playing the game (with the optimal strategy) to go from \$1 to \$1,000,000.

In information theory, the expression

$$c = 1 + p \log p + q \log q$$

is known as the *channel capacity* of a binary symmetric channel. The channel capacity measures the rate at which information can be reliably sent over a noisy channel.

In a *binary symmetric channel* (BSC), each binary symbol, 0 and 1, is sent accurately over a channel with probability p and inaccurately with probability $q = 1 - p$. See Figure 17.1.

The *capacity* $c(p)$ of a BSC is defined as

$$c(p) = 1 - H(p) = 1 + p \log p + q \log q.$$

The graph of the capacity function $c(p)$ is shown in Figure 17.2. Observe that the capacity function is an "upside-down" version of the entropy function (compare with Figure 16.2).

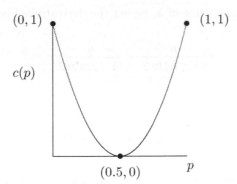

FIGURE 17.2: The capacity function $c(p)$.

We see from Figure 17.2 that $c(0.5) = 0$, i.e., a completely random channel cannot convey any information, and $c(0) = c(1) = 1$, i.e., a channel that is completely accurate or completely inaccurate can convey information perfectly.

Shannon's second theorem of information theory (which we will see in Chapter 18) states that information can be sent (using an error-correcting code) over a BSC with arbitrarily high accuracy at any rate below the channel capacity.

We have found the optimal fraction of our amount to bet in the coin-tossing game. Let's analyze the game further to find the failure threshold, that is, the smallest fraction which, when bet consistently, will result in long-term ruin.

Suppose that we start with $1 and on each day we bet some fraction λ, $0 < \lambda < 1$, of our current holdings. Each investment has an even money payoff, with probability of success p and failure $q = 1 - p$. Let X_n be our amount after the nth coin toss.

Let's suppose that in n coin tosses we obtain s heads (successes) and f tails (failures), where $s + f = n$. Then

$$X_n = (1 + \lambda)^s (1 - \lambda)^f.$$

The expected values of s and f are np and nq, respectively. Hence, the expected value of X_n is

$$(1 + \lambda)^{pn} (1 - \lambda)^{qn} = 2^{g(\lambda)n},$$

where

$$g(\lambda) = p \log(1 + \lambda) + q \log(1 - \lambda).$$

Logarithms are base 2. We call $g(\lambda)$ the *growth rate function*.

To find the optimal value of λ, we set the derivative g' equal to 0. Thus

$$g' = \frac{p}{(1+\lambda)\ln 2} - \frac{q}{(1+\lambda)\ln 2} = 0,$$

which implies that

$$p(1-\lambda) - q(1+\lambda) = 0.$$

Hence, the optimal value of λ, call it λ^*, is given by $\lambda^* = \lambda^*(p+q) = p-q$ (we already knew this). Furthermore,

$$g'' = \frac{-p}{(1+\lambda)^2 \ln 2} + \frac{-q}{(1-\lambda)^2 \ln 2} < 0.$$

We find that g is a concave downward function of λ, with one maximum and one point, λ_0, where the function crosses the horizontal axis (the *critical value*). See Figure 17.3.

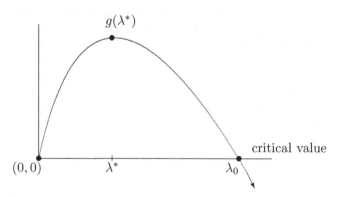

FIGURE 17.3: The growth rate function $g(\lambda)$.

In our original example, $p = 0.53$, $q = 0.47$, and $\lambda^* = p-q = 0.06$. As we have said, the best strategy is to consistently bet 6 per cent of our current amount. In this case, the coefficient of growth is approximately 0.00259841. The critical value is $\lambda_0 \doteq 0.12$. If we consistently wager more than λ_0, then we are on the road to ruin.

Exercises

◇1. In the coin-tossing game, given a coin with probability 0.51 of landing heads, what is the optimal fraction to bet? How long would it take (on average) to go from \$1 to \$1,000,000, following the optimal strategy?

◇2. In the game of the previous exercise, what is the smallest fraction that leads to ruin?

◇3. In the coin-tossing game, given a coin with probability 0.9 of landing heads, what is the optimal fraction to bet? How long would it take (on average) to go from \$1 to \$1,000,000, following the optimal strategy?

◇4. In the game of the previous exercise, what is the smallest fraction that leads to ruin?

5. Suppose that there are three outcomes, A, B, and C, with probabilities p, q, and r, respectively. Suppose that these outcomes give payoffs of x, y, and z, per unit bet, respectively. How much of our current amount should be bet on each outcome?

†⋆6. Generalize the result of the previous exercise.

7. What is the capacity of a BSC with $p = 0.75$? What is the capacity if $p = 0.25$?

8. If three symbols are sent over a BSC with $p = 0.9$, what is the probability that two or three of the symbols are received correctly?

Chapter 18

Shannon's Theorems

The average length of a source code is at least equal to the entropy of the source.

Information can be reliably sent over a noisy channel at any rate below the channel capacity.

A *code* C for a source with n states is a sequence $w_1, \ldots, w_n$ of binary strings, none the prefix of another. The w_i are the *words* of the code. We may now write our source as

$$
\begin{pmatrix}
x_1 & x_2 & \cdots & x_n \\
p_1 & p_2 & \cdots & p_n \\
w_1 & w_2 & \cdots & w_n
\end{pmatrix}.
$$

The *length* l_i of a code word w_i is the number of bits in w_i.

The following pivotal result is credited to Leon G. Kraft Jr.

Theorem 18.1 (Kraft's Inequality, 1949). A source with n states has a code with word lengths $l_1, \ldots, l_n$ if and only if

$$
\sum_{i=1}^{n} 2^{-l_i} \leq 1.
$$

Proof. Let l be the maximum of the l_i. Then a word of length l_i in the code prevents 2^{l-l_i} binary strings of length l from being code words (because of the "no prefix" rule). Therefore, the encoding is possible if and only if $\sum_{i=1}^{n} 2^{l-l_i} \leq 2^l$, i.e., $\sum_{i=1}^{n} 2^{-l_i} \leq 1$. ∎

Example 18.2. Verify Kraft's inequality for the code

$$
\begin{pmatrix}
x_1 & x_2 & x_3 \\
1/2 & 1/4 & 1/4 \\
0 & 10 & 11
\end{pmatrix}.
$$

Solution: Note that none of the code words is a prefix of another (so the code is legitimate). We have

$$\sum_{i=1}^{n} 2^{-l_i} = \frac{1}{2} + \frac{1}{4} + \frac{1}{4} = 1 \leq 1.$$

∎

The *average length* of the code C is $\bar{l} = \sum_{i=1}^{n} p_i l_i$. This quantity is the average number of code symbols per source symbol.

Example 18.3. Find the average length of the code of the previous example.

Solution: We have

$$\bar{l} = \frac{1}{2} \cdot 1 + \frac{1}{4} \cdot 2 + \frac{1}{4} \cdot 2 = \frac{3}{2}.$$

∎

Shannon's seminal result in noiseless coding theory says that the average length of a code is at least equal to the entropy of the source.

Theorem 18.4 (Shannon's First Theorem, 1948). *The average length $\bar{l}$ of a code for a source S satisfies $\bar{l} \geq H(S)$.*

Proof.

$$
\begin{aligned}
\bar{l} &= \sum p_i l_i \\
&\geq \sum p_i l_i + \log\left(\sum 2^{-l_i}\right) \quad \text{(by Theorem 18.1)} \\
&= \sum p_i l_i + \sum p_i \log\left(\sum 2^{-l_i}\right) \\
&= -\sum p_i \log\left(\frac{2^{-l_i}}{\sum 2^{-l_i}}\right) \\
&\geq -\sum p_i \log p_i \quad \text{(by Lemma 16.4)} \\
&= H(S)
\end{aligned}
$$

The second inequality follows from Lemma 16.4, since

$$\sum_{i=1}^{n} \frac{2^{-l_i}}{\sum_{i=1}^{n} 2^{-l_i}} = 1.$$

∎

Example 18.5. Verify Shannon's first theorem for the code of the previous example.

Solution: We calculated that $\bar{l} = 3/2$. The entropy of the code is

$$H(S) = -\frac{1}{2}\log\frac{1}{2} - \frac{1}{4}\log\frac{1}{4} - \frac{1}{4}\log\frac{1}{4} = \frac{3}{2}.$$

So, in this case, the average length and the entropy are equal. ■

As we have said, Shannon's theorem asserts that the entropy of a source is a lower bound for the average number of code symbols needed to encode each source symbol. In fact, the lower bound is nearly achievable.

Theorem 18.6. Given a source S, there exists a code for S with average length less than $H(S) + 1$.

Proof. Suppose that the source has n symbols which occur with probabilities $p_1, \ldots, p_n$. For $1 \le i \le n$, define l_i to be the integer for which

$$-\log p_i \le l_i < -\log p_i + 1.$$

Then

$$\sum_{i=1}^{n} 2^{-l_i} \le \sum_{i=1}^{n} 2^{\log p_i} = \sum_{i=1}^{n} p_i = 1.$$

By the Kraft inequality, we can encode S with strings of lengths $l_1, \ldots, l_n$. Moreover, the average length of the code, $\bar{l}$, satisfies

$$\bar{l} < \sum_{i=1}^{n} (-\log p_i + 1)p_i = H(S) + 1.$$

■

Example 18.7. Let

$$S = \begin{pmatrix} x_1 & x_2 & x_3 & x_4 & x_5 & x_6 & x_7 & x_8 \\ 0.2 & 0.2 & 0.1 & 0.1 & 0.1 & 0.1 & 0.1 & 0.1 \end{pmatrix}.$$

What is the entropy of this source? Find a code for this source that has average length within 1 bit of the entropy.

Solution: We have $H(S) = 2(-0.2\log 0.2) + 6(-0.1\log 0.1) \doteq 2.9$. According to Theorem 18.4, we can find a code for S within 1 bit of this quantity. Following the proof of the theorem, we take the code lengths to be $l_1, l_2, l_3, l_4, l_5, l_6, l_7, l_8$, where

$$-\log 0.2 \le l_1, l_2 < -\log 0.2 + 1,$$

$$-\log 0.1 \le l_3, l_4, l_5, l_6, l_7, l_8 < -\log 0.1 + 1.$$

Thus $l_1 = l_2 = 3$ and $l_3 = l_4 = l_5 = l_6 = l_7 = l_8 = 4$. These choices give an average code length of $2(0.2 \cdot 3) + 6(0.1 \cdot 4) = 3.6$. In fact, a suitable code is given by

$$\begin{pmatrix} x_1 & x_2 & x_3 & x_4 & x_5 & x_6 & x_7 & x_8 \\ 0.2 & 0.2 & 0.1 & 0.1 & 0.1 & 0.1 & 0.1 & 0.1 \\ 100 & 101 & 0000 & 0001 & 0010 & 0011 & 0100 & 0101 \end{pmatrix}.$$

■

For sources X and Y, with

$$X = \begin{pmatrix} x_1 & x_2 & \cdots & x_m \\ p_1 & p_2 & \cdots & p_m \end{pmatrix}$$

and

$$Y = \begin{pmatrix} y_1 & y_2 & \cdots & y_n \\ q_1 & q_2 & \cdots & q_n \end{pmatrix},$$

we define the *product source* XY as

$$XY = \begin{pmatrix} (x_i, y_j) \\ p_i q_j \end{pmatrix},$$

where $i = 1, \ldots, m$ and $j = 1, \ldots, n$.

Proposition 18.8. Given any two sources X and Y, we have

$$H(XY) = H(X) + H(Y).$$

Proof.

$$\begin{aligned} H(XY) &= -\sum_{i=1}^{m} \sum_{j=1}^{n} p_i q_j \log p_i q_j \\ &= -\sum_{i=1}^{m} \sum_{j=1}^{n} p_i q_j \log p_i - \sum_{i=1}^{m} \sum_{j=1}^{n} p_i q_j \log q_j \\ &= -\sum_{i=1}^{m} p_i \log p_i \sum_{j=1}^{n} q_j - \sum_{i=1}^{m} p_i \sum_{j=1}^{n} q_j \log q_j \\ &= H(X) + H(Y) \end{aligned}$$

■

The nth product of a source S with itself, written S^n, is called the nth *extension* of S.

Corollary 18.9. Given a source S and any integer $n \geq 1$, we have

$$H(S^n) = nH(S).$$

Corollary 18.10. There is a sequence of codes for S^n with average lengths $\bar{l}_n$ satisfying

$$\lim_{n \to \infty} \frac{\bar{l}_n}{n} = H(S).$$

Proof. By Theorems 18.4 and 18.6, there exists a code with

$$H(S^n) \leq \bar{l}_n < H(S^n) + 1.$$

By Corollary 18.9, we have $nH(S) \leq \bar{l}_n < nH(S) + 1$, i.e.,

$$H(S) \leq \bar{l}_n/n < H(S) + 1/n.$$

Letting $n \to \infty$, the conclusion follows. ∎

This result says that, in the limit, the entropy of a source is equal to the number of bits needed to encode its states.

A *channel* (X, Y) consists of an *input alphabet* $X = \{x_1, x_2, \ldots, x_m\}$, an *output alphabet* $Y = \{y_1, y_2, \ldots, y_n\}$, and conditional probabilities $p_{ij} = p(y_j|x_i)$, for $1 \leq i \leq m$, $1 \leq j \leq n$. The conditional probability $p(y_j|x_i)$ is the probability that the output symbol y_j is received when the input symbol x_i is sent. We represent the channel as

$$
\begin{array}{cc|ccc}
 & & Y & & \\
 & & y_1 & \cdots & y_n \\
 & & q_1 & \cdots & q_n \\
\hline
X & x_1 & p_1 & & \\
 & \vdots & \vdots & & p(y_j|x_i) \\
 & x_m & p_m & & \\
\end{array}
$$

or by a matrix of conditional probabilities:

$$
\begin{bmatrix}
p_{11} & \cdots & p_{1n} \\
\vdots & & \vdots \\
p_{m1} & \cdots & p_{mn}
\end{bmatrix}.
$$

What we have defined as a channel is usually called a *discrete memoryless channel* (DMC). "Memoryless" means that the appearance of a symbol has

noise

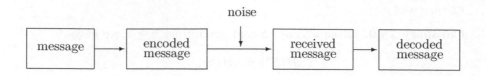

FIGURE 18.1: A communication model.

no bearing on the probability distribution of future symbols. The communication model in Figure 18.1 includes a channel of information from an encoded message to a received message.

Given a channel (X, Y), various entropies are defined:

1. *Input entropy*:
$$H(X) = -\sum_{i=1}^{m} p_i \log p_i;$$

2. *Output entropy*:
$$H(Y) = -\sum_{j=1}^{n} q_j \log q_j;$$

3. *Conditional entropy* or *equivocation*:
$$H(X|Y) = -\sum_{i=1}^{m}\sum_{j=1}^{n} p(x_i, y_j) \log p(x_i|y_j)$$

and
$$H(Y|X) = -\sum_{i=1}^{m}\sum_{j=1}^{n} p(x_i, y_j) \log p(y_j|x_i);$$

4. *Total entropy*:
$$H(X, Y) = -\sum_{i=1}^{m}\sum_{j=1}^{n} p(x_i, y_j) \log p(x_i, y_j);$$

5. *Mutual information*:
$$I(X, Y) = H(X) - H(X|Y).$$

Figure 18.2 indicates the relationships among the various entropies. By

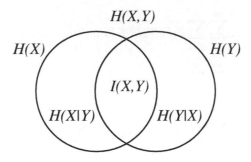

FIGURE 18.2: The entropies associated with a channel.

analogy with *a priori* and *a posteriori* probabilities, $H(X)$ and $H(X|Y)$ are called *a priori* entropy and *a posteriori* entropy, respectively.

We now explain the definition of $H(X|Y)$. Suppose that Y is observed to equal y_j. Then the amount of uncertainty in X is

$$H(X|Y = y_j) = -\sum_{i=1}^{m} p(x_i|y_j) \log p(x_i|y_j).$$

Hence the average amount of uncertainty that remains in X when Y is observed is

$$
\begin{aligned}
H(X|Y) &= \sum_{j=1}^{n} H(X|Y = y_j)q_j \\
&= -\sum_{i=1}^{m}\sum_{j=1}^{n} p(x_i|y_j)q_j \log p(x_i|y_j) \\
&= -\sum_{i=1}^{m}\sum_{j=1}^{n} p(x_i, y_j) \log p(x_i|y_j).
\end{aligned}
$$

Therefore, we may think of $H(X|Y)$ as the amount of information lost in the channel. This is consistent with the following theorem.

Theorem 18.11. Given any channel (X, Y), we have

$$H(X, Y) = H(Y) + H(X|Y).$$

Proof.

$$H(X,Y) = -\sum_{i=1}^{m}\sum_{j=1}^{n} p(x_i, y_j) \log p(x_i, y_j)$$

$$= -\sum_{i=1}^{m}\sum_{j=1}^{n} p(x_i, y_j)[\log q_j + \log p(x_i|y_j)]$$

$$= -\sum_{i=1}^{m}\sum_{j=1}^{n} p(x_i, y_j)\log q_j - \sum_{i=1}^{m}\sum_{j=1}^{n} p(x_i, y_j)\log p(x_i|y_j)$$

$$= -\sum_{j=1}^{n} q_j \log q_j + H(X|Y)$$

$$= H(Y) + H(X|Y)$$

$\blacksquare$

Similarly, $H(X,Y) = H(X) + H(Y|X)$. It follows that $I(X,Y) = H(Y) - H(Y|X)$.

Theorem 18.12. Given any channel (X,Y), we have

$$H(X,Y) \le H(X) + H(Y).$$

Equality holds if and only if X and Y are independent.

Proof.

$$H(X,Y) = -\sum_{i=1}^{m}\sum_{j=1}^{n} p(x_i, y_j)\log p(x_i, y_j)$$

$$\le -\sum_{i=1}^{m}\sum_{j=1}^{n} p(x_i, y_j)\log p_i q_j \quad \text{(by Lemma 16.4)}$$

$$= -\sum_{i=1}^{m}\sum_{j=1}^{n} p(x_i, y_j)\log p_i - \sum_{i=1}^{m}\sum_{j=1}^{n} p(x_i, y_j)\log q_j$$

$$= H(X) + H(Y)$$

According to Lemma 16.4, equality occurs if and only if $p(x_i, y_j) = p_i q_j$ for all i and j, i.e., when X and Y are independent. $\blacksquare$

Corollary 18.13. Given any channel (X,Y), the following inequalities hold:

1. $0 \leq H(X|Y) \leq H(X)$;

2. $0 \leq H(Y|X) \leq H(Y)$;

3. $0 \leq I(X, Y)$.

The *binary symmetric channel* (BSC) is the channel in which X and Y each have two symbols, 0 and 1, and the probability that a symbol is sent accurately over the channel is independent of whether the symbol is a 0 or a 1. For definiteness, let us suppose that each symbol is sent accurately with probability p and inaccurately with probability $q = 1 - p$. See Figure 17.1.

The BSC is represented by the matrix

$$\begin{bmatrix} p & q \\ q & p \end{bmatrix}.$$

The matrix representation of BSC^n is the nth Kronecker power of the BSC matrix.

Given a channel from a source X to a source Y, we always have the inequality

$$H(Y) \geq H(X).$$

Intuitively, the channel only adds uncertainty to a source. More formally, we can prove the inequality by noting that $px + qy \leq y$, where $x \leq y$.

The *capacity* c of a channel is defined as the maximum mutual information of the channel:

$$c = \max_{\{p_i\}} I(X, Y).$$

We denote by $c(p)$ the capacity of the binary symmetric channel $\text{BSC}(p)$.

Theorem 18.14. The capacity $c(p)$ is given by the formula

$$c(p) = 1 + p \log p + q \log q = 1 - H(p).$$

Proof. Suppose that 0 and 1 are transmitted with probabilities x and y, respectively. We have

$$I(Y, X) = H(Y) - H(Y|X) = H(Y) - H(p) \leq 1 - H(p),$$

with equality if $x = y = 1/2$. Hence

$$c(p) = 1 - H(p) = 1 + p \log p + q \log q.$$

∎

The capacity function $c(p)$ is shown in Figure 17.2.

Seeing the entropy function $H(p)$ occur in many of the above theorems, and seeing the proofs of the theorems work out so perfectly, we begin to appreciate that the entropy function, like the exponential and trigonometric functions, is a fundamental function of mathematics.

The rate of a code is a measure of the number of code words versus the number of bits in the code words. Given a code C with all code words of length n, we define the *rate* of C to be

$$r(C) = \frac{\log_2 |C|}{n}.$$

Notice that $r(C) = 1$ if and only if C consists of all binary strings of length n. Otherwise, $r(C)$ is strictly less than 1.

Shannon's second theorem asserts that we can send information over a noisy channel with an arbitrarily high degree of accuracy, as long as the rate is less than the channel capacity.

Theorem 18.15 (Shannon's Second Theorem, 1948). Consider the binary symmetric channel BSC with probability of error $p < 1/2$ and capacity $c = 1 - H(p)$. Let $R < c$ and $\epsilon > 0$. For sufficiently large n, there exists a subset of $M \geq 2^{Rn}$ code words (to represent M equally probable messages) from the set of 2^n possible inputs to the BSC^n such that the probability of error (per word) is less than ϵ.

The code guaranteed by Shannon's second theorem has rate

$$\frac{\log_2 M}{n} \geq \frac{\log_2 2^{Rn}}{n} = R.$$

Thus it is possible, by choosing n sufficiently large, to reduce the maximum probability of error to an amount as low as desired while at the same time maintaining the transmission rate near the channel capacity.

Lemma 18.16. If n is a positive integer and $0 < x < 1/2$, then

$$\sum_{k=0}^{\lfloor nx \rfloor} \binom{n}{k} < 2^{nH(x)}.$$

Proof. Let $y = 1 - x$. Then

$$1 = (x + y)^n$$

$$= \sum_{k=0}^{n} \binom{n}{k} x^k y^{n-k}$$

$$> \sum_{k=0}^{\lfloor nx \rfloor} \binom{n}{k} (x/y)^k y^n$$

$$> \sum_{k=0}^{\lfloor nx \rfloor} \binom{n}{k} (x/y)^{nx} y^n$$

$$= \sum_{k=0}^{\lfloor nx \rfloor} \binom{n}{k} x^{nx} y^{ny}.$$

Therefore

$$\sum_{k=0}^{\lfloor nx \rfloor} \binom{n}{k} < x^{-nx} y^{-ny} = 2^{nH(x)}.$$

∎

Proof of theorem. The proof uses the probabilistic method, in which the existence of a desired object (a good code) is established by showing that it exists with positive probability.

We establish some technical preliminaries. Choose R' with $R < R' < c$. Let $\delta = \epsilon/2$. Choose Δ so that

$$R' < 1 - H(p_\Delta) = c(p_\Delta) < c,$$

where $p_\Delta = p + \Delta$. See Figure 18.3.

The *Hamming distance* between two code words, named after Richard Hamming (1915–1998), is the number of coordinates in which the words differ. The Hamming distance of a code is the minimum Hamming distance between code words. Assume that the channel is a BSC^n with probability of error p, with n to be determined. Suppose that α is transmitted and β is received. The expected Hamming distance between α and β is np. Consider a sphere T of radius np_Δ about β. Our decision procedure is as follows: if there is a unique word in T, then we accept it. If there is no code word in T, or more than one code word, then we concede an error.

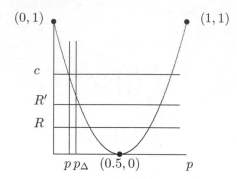

FIGURE 18.3: Ingredients for Shannon's second theorem.

The probability of error is

$$
\begin{aligned}
\Pr(\text{error}) \;=\;& \Pr(\alpha \notin T) + \Pr(\alpha \in T)\Pr(\alpha' \in T : \alpha' \neq \alpha) \\
\leq\;& \Pr(\alpha \notin T) + \Pr(\alpha' \in T : \alpha' \neq \alpha) \\
\leq\;& \Pr(\alpha \notin T) + \sum_{\alpha' \neq \alpha} \Pr(\alpha' \in T).
\end{aligned}
$$

The second inequality is due to the subadditivity of probabilities (Bonferroni inequalities).

It follows from the law of large numbers that, given Δ and δ, there exists n_0 such that

$$
\Pr\left(\frac{|X - np|}{n} > \Delta\right) < \delta,
$$

for $n \geq n_0$. Hence, the probability that the number of errors, X, exceeds the expected number of errors, np, by more than $n\Delta$ is less than δ. Therefore, we may make the first term arbitrarily small (less than δ).

This argument for the first term is independent of the M code words chosen. However, the argument for the second term, $\sum_{\alpha' \neq \alpha} \Pr(\alpha' \in T)$, is not. Choose M with $2^{nR} \leq M \leq 2^{nR'}$. Suppose that M words are selected randomly from the 2^n possible words. There are 2^{nM} possible codes, each selected with probability 2^{-nM}. Thus

$$
\begin{aligned}
\overline{\Pr}(\text{error}) \;<\;& \delta + (M-1)\overline{\Pr}(\alpha' \in T) \quad (\alpha' \neq \alpha) \\
\leq\;& \delta + M\overline{\Pr}(\alpha' \in T),
\end{aligned}
$$

where $\overline{\Pr}$ denotes an average probability over all 2^{nM} codes.

Now $\Pr(\alpha' \in T) = |T|/2^n$, where $|T| = \sum_{k=0}^{\lfloor np_\Delta \rfloor} \binom{n}{k}$. Therefore, by Lemma 18.16, we have

$$
\begin{aligned}
\overline{\Pr}(\text{error}) \;&<\; \delta + M2^{-n}2^{nH(p_\Delta)} \\
&\leq\; \delta + 2^{nR'-n+nH(p_\Delta)} \\
&=\; \delta + 2^{n(R'-1+H(p_\Delta))} \\
&=\; \delta + 2^{n(R'-c(p_\Delta))}.
\end{aligned}
$$

The second term may be made less than δ for $n \geq n_1$. Take $n = \max(n_0, n_1)$. This makes the average probability of error, over all codes of size M, less than ϵ. Therefore, there *exists* such a code. ∎

According to Shannon's second theorem, there is a trade-off between information rate and error-correcting capability of a code. The main goal in the area of error-correcting codes is to produce codes with both high information rate and high error-correcting capability.

Exercises

1. Encode a, b, c, d, e, f, g with binary strings (no string an extension of another) with lengths 4, 4, 3, 3, 3, 2, 2.

2. Describe how to construct a code for n symbols with word lengths 1, 2, ... n. If the corresponding words occur with probabilities $1/2$, $1/4$, ..., $1/2^n$, respectively, what is the average length of the code?

3. Find a binary code for S in Exercise 16.8 so that the average word length is within one bit of $H(S)$.

4. For the channel (X, Y) given below, find $H(X)$, $H(Y)$, $H(X,Y)$, $H(X|Y)$, $H(Y|X)$, and $I(X,Y)$.

			Y	
			y_1	y_2
			5/8	3/8
X	x_1	3/4	1/4	3/4
	x_2	1/4	1/8	7/8

5. Find the rate of the code $\{011, 101, 110\}$.

6. In a *triplicate code*, each bit is sent over a BSC three times. What is the rate of the triplicate code? If $p = 0.9$, what is the per information bit error rate of the triplicate code?

7. Find a binary code with words of length 7 and rate 4/7.

†8. Show that if a binary code C consisting of strings of length n has minimum Hamming distance d, then

$$|C| \le \frac{2^n}{\sum_{e=0}^{\lfloor (d-1)/2 \rfloor} \binom{n}{e}}.$$

This is known as the *Hamming bound* or *sphere-packing bound*. If equality occurs, then we say that C is a *perfect binary code*.

9. Give an example of a perfect binary code of length 7.

⋆10. Suppose that k binary strings are randomly and independently chosen from the set of binary strings of length n. If the chosen strings are taken to be the words of a code, what is the expected minimum Hamming distance of the code?

Part VII

Games

Chapter 19

A Little Graph Theory Background

In any group of six people, there are three all of whom know each other or three none of whom know each other.

A *graph* G consists of a set V of *vertices* (also called *points* or *nodes*) and a set E of *edges* (also called *lines* or *arcs*) which are unordered pairs of vertices. We sometimes denote an edge $\{x, y\}$ as xy. In a drawing of a graph, two vertices x and y are joined by a line if and only if $\{x, y\} \in E$. If two vertices are joined by a line, they are *adjacent*; otherwise, they are *nonadjacent*. If $|V| = p$ and $|E| = q$, then we say that G has *order* p and *size* q. Some good references on graph theory are [Bol79], [CL96], [CZ05], [Har69], and [Wes95].

The *complete graph* K_n consists of n vertices and all $\binom{n}{2}$ possible edges. The *complete bipartite graph* $K_{m,n}$ consists of a set A of m points, a set B of n points, and all the mn edges between A and B. The *cycle* C_n consists of n vertices connected by n edges in a circuit. The *path* P_n is C_n minus an edge. Figure 19.1 illustrates some of these graphs (without vertex labels).

Given a vertex v of a graph, we denote by $\delta(v)$ the *degree* of v, that is, the number of vertices adjacent to v. If $\delta(v)$ is the same for all vertices, then the graph is *regular* of *degree* $\delta(v)$. Complete graphs and cycles are regular. For example, K_3, $K_{2,2}$, and C_4 are each regular of degree 2.

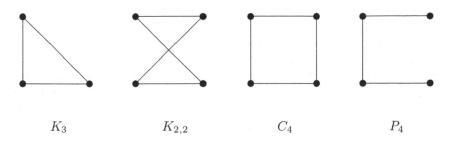

$$K_3 \qquad\qquad K_{2,2} \qquad\qquad C_4 \qquad\qquad P_4$$

FIGURE 19.1: Examples of graphs.

155

Two graphs are isomorphic if the vertices of one graph can be relabeled to create the second graph. For example, $K_{2,2}$ and C_4 are isomorphic.

The *graph complement* G^c of a graph G consists of the vertices of G and the non-edges of G. For example, P_4 is isomorphic to its own complement.

A formula for the number of nonisomorphic graphs of given order is derived in [Eri96]. As an example, there are 12,005,168 nonisomorphic graphs of order 10.

We will prove a few simple but important facts about graphs. The pigeonhole principle will be a useful tool.

Proposition 19.1 (Pigeonhole Principle). Given any function $f : X \to Y$, where $|X| = |Y| + 1$, we have $f(x_1) = f(x_2)$ for some distinct x_1, $x_2 \in X$.

Proof. The result is so obvious it hardly needs proving, but we offer a proof by contradiction. If the result were not true, then the inverse image under f of each element in Y would consist of at most one element of X. But this implies that the cardinality of X is at most equal to the cardinality of Y, which is false. This contradiction establishes that the result is true. ∎

The pigeonhole principle is often paraphrased as follows: If $n+1$ objects are placed in n pigeonholes, then at least one pigeonhole must contain at least two objects. Johann Peter Gustav Lejeune Dirichlet (1805–1859) was the first mathematician to explicitly use the pigeonhole principle in proofs. He referred to it as the "drawer principle."

We consider an "extremal property" of graphs. How many edges are possible in a triangle-free graph with $2n$ vertices? The complete bipartite graph $K_{n,n}$ has n^2 edges and no triangle. In fact, n^2 is the maximum number of edges in a triangle-free graph with $2n$ vertices.

Theorem 19.2 (W. Mantel, 1907). A graph with $2n$ vertices and $n^2 + 1$ edges must contain a triangle.

Proof 1 (Mathematical Induction). Let G be a graph with $2n$ vertices and $n^2 + 1$ edges. If $n = 1$, then G cannot have $n^2 + 1$ edges; hence the result is vacuously true. Assuming the result for n, we consider a graph G with $2(n + 1)$ vertices and $(n + 1)^2 + 1$ edges. Let x and y be adjacent vertices in G and let H be G minus x and y and any edges from x and y. If H, a graph with $2n$ vertices, has more than n^2 edges, then the result holds by the induction hypothesis. So suppose that H has at most n^2 edges and hence there exist at least $2n + 1$ edges joining x and y to vertices in H. By the pigeonhole principle, there exists a vertex z in H that is adjacent to both x and y. Therefore G contains a triangle xyz. ∎

We give a second proof that doesn't use the pigeonhole principle. Denote by $\alpha(G)$ the maximum number of pairwise nonadjacent vertices in G. This is called the *independence number* of G.

Proof 2 (AM–GM Inequality). Suppose that the graph G has no triangles. Let $\alpha = \alpha(G)$ and let I be an independent set of α vertices. Let $\beta = 2n - \alpha$ (the cardinality of $V - I$). We make two simple observations: (1) Every edge of G has at least one vertex in $V - I$ (since I is independent). (2) Every vertex of G has degree at most α (since G is triangle-free). By (1), (2), and the arithmetic mean-geometric mean (AM–GM) inequality,

$$|E| \le \sum_{v \in V-I} \delta(v) \le \alpha\beta \le \left(\frac{\alpha + \beta}{2}\right) = n^2.$$

This contradicts the assumption that $|E| = n^2 + 1$. Therefore G contains a triangle. ∎

Mantel's theorem is a special case of an important theorem of Pál Turán (1910–1976). A *subgraph* of a graph G is a graph obtained from G by possibly removing some vertices and edges.

Theorem 19.3 (P. Turán, 1941). In a graph with n vertices containing no complete subgraph K_m, the number of edges is at most

$$\frac{(m-2)n}{2(m-1)}.$$

For a proof, see [Bol79].

Now let's prove the result from the chapter teaser: In any group of six people, there are three all of whom know each other or three none of whom know each other.

The six people and the relations between each pair of them ("know each other" or "don't know each other") can be represented via the complete graph K_6. We think of the people as the vertices of the graph and the relations as the edges.

A *coloring* of the set of edges of a graph G is an assignment of a *color* to each edge of G. If all the edges of G have the same color, then G is *monochromatic*. We want to show that if each edge of K_6 is colored either green or red, then there is a monochromatic subgraph K_3 (a triangle). The coloring may be done in an arbitrary manner. In fact, since K_6 has $\binom{6}{2} = 15$ edges, there are $2^{15} = 32{,}768$ possible green–red colorings of the edges of K_6. We claim that every coloring has a monochromatic subgraph K_3.

Assume that the edges of K_6 are colored using green and red colors, and choose any vertex v. By the pigeonhole principle, at least three of the five edges from v are the same color. Without loss of generality, suppose that v is joined by green edges to vertices x, y, z. If any of the edges xy, yz, or xz is green, then there is a green triangle (vxy, vyz, or vxz). On the other hand, if each of these edges is red, then xyz is a red triangle.

In fact, six is the smallest number of people that force the property of having three all of whom know each other or all of whom don't know each other. This is because there is a 2-coloring of the edges of K_5 without a monochromatic triangle (see Exercises).

In 1930 Frank Ramsey (1903–1930) proved that for every positive integer m, there exists an integer n such that every 2-coloring of the edges of K_n has a monochromatic subgraph K_m.

Theorem 19.4 (Ramsey's Theorem, 1930). Given integers $a, b \geq 2$, there exists a least integer $R(a, b)$ with the following property: Every green–red coloring of the edges of the complete graph on $R(a, b)$ vertices yields a subgraph K_a all of whose edges are green or a subgraph K_b all of whose edges are red. Furthermore

$$R(a, b) \leq R(a - 1, b) + R(a, b - 1), \quad a, b \geq 3.$$

Proof (Induction on a and b). The values $R(a, 2) = a$ and $R(2, b) = b$ are trivial (why?). These values are the basis of the induction. Assume that $R(a - 1, b)$ and $R(a, b - 1)$ exist; we will show that $R(a, b)$ exists. Let G be the complete graph on $R(a - 1, b) + R(a, b - 1)$ vertices, and let v be an arbitrary vertex of G. By the pigeonhole principle, at least $R(a-1, b)$ green edges or at least $R(a, b - 1)$ red edges are incident with v. Without loss of generality, suppose that v is joined by green edges to a complete subgraph on $R(a - 1, b)$ vertices. By definition of $R(a - 1, b)$, this subgraph must contain a green subgraph K_{a-1} or a red subgraph K_b. In the former case, the green subgraph K_{a-1}, v, and all the edges between the two constitute a green subgraph K_a. We have shown that G contains a green subgraph K_a or a red subgraph K_b. Therefore, $R(a, b)$ exists and satisfies the inequality

$$R(a, b) \leq R(a - 1, b) + R(a, b - 1).$$

∎

The integers $R(a, b)$ are called *Ramsey numbers*. Very few Ramsey numbers have been calculated.

We mentioned in the proof that $R(a, 2) = a$ for all $a \geq 2$. Also, by symmetry, $R(a, b) = R(b, a)$ for all $a, b \geq 2$. Furthermore, the inequality of the theorem is very important in finding bounds for Ramsey numbers.

The values $R(a, a)$ are called *diagonal Ramsey numbers*. From the result of the chapter introduction and one of the exercises, we know one diagonal Ramsey number: $R(3, 3) = 6$. The only other nontrivial diagonal Ramsey number known is $R(4, 4)$, which we will address shortly.

Ramsey's theorem has a straightforward generalization to edge-coloring with an arbitrary number of colors.

Theorem 19.5 (Ramsey's Theorem for Arbitrarily Many Colors). For any integer $c \geq 2$ and integers $a_1, \ldots, a_c \geq 2$, there exists a least integer $R(a_1, \ldots, a_c)$ with the following property: If the edges of the complete graph on $R(a_1, \ldots, a_c)$ vertices are colored with colors $\alpha_1, \ldots, \alpha_c$, then for some i there exists a complete subgraph on a_i vertices all of whose edges are color α_i.

Proof. The case $c = 2$ is covered by our previous version of Ramsey's theorem. Suppose that $R(a_1, \ldots, a_{c-1})$ exists for all $a_1, \ldots, a_{c-1} \geq 2$. We claim that $R(a_1, \ldots, a_c)$ exists and satisfies

$$R(a_1, \ldots, a_c) \leq R(R(a_1, \ldots, a_{c-1}), a_c).$$

A c-coloring of the complete graph on $R(R(a_1, \ldots, a_{c-1}), a_c)$ vertices may be regarded as a 2-coloring with colors $\{\alpha_1, \ldots, \alpha_{c-1}\}$ and α_c. Such a coloring contains a complete graph on a_c vertices colored α_c or a $(c-1)$-colored complete graph on $R(a_1, \ldots, a_{c-1})$ vertices, using colors $\alpha_1, \ldots, \alpha_{c-1}$, in which case the induction hypothesis holds. In either case, we obtain a complete monochromatic subgraph on the appropriate number of vertices. ∎

The c-color Ramsey numbers $R(a_1, \ldots, a_c)$ satisfy certain trivial relations, e.g., they are symmetric in the c variables. Furthermore

$$R(a_1, \ldots, a_{c-1}, 2) = R(a_1, \ldots, a_{c-1}), \quad a_i \geq 2,$$

because there is either an edge colored α_c or else all edges are colored from the set $\{\alpha_1, \ldots, \alpha_{c-1}\}$.

We know that $R(3, 3) = 6$. Let's evaluate the Ramsey number $R(3, 4)$. To obtain an upper bound, we use the inequality of Theorem 19.4:

$$R(3, 4) \leq R(3, 3) + R(2, 4) = 6 + 4 = 10.$$

In fact, $R(3, 4) = 9$. For suppose that there is a green–red coloring of K_9 with no green subgraph K_3 and no red subgraph K_4. Since $R(2, 4) = 4$ and $R(3, 3) = 6$, each vertex of the graph K_9 must be incident with exactly three green edges and five red edges. But this implies that the sum of the degrees of the vertices of the green subgraph is $9 \cdot 3 = 27$, contradicting the fact that the sum of degrees is always even (it is twice the number of edges). Hence $R(3, 4) \leq 9$. In the exercises, you are asked to furnish a 2-coloring of K_8 containing no green subgraph K_3 and no red subgraph K_4, thereby proving that $R(3, 4) = 9$.

Next we determine the Ramsey number $R(4, 4)$. The inequality of Theorem 19.4 yields the upper bound

$$R(4, 4) \leq R(4, 3) + R(3, 4) = 9 + 9 = 18.$$

In fact, $R(4, 4) = 18$. To prove this, it suffices to show a green–red coloring of K_{17} containing no monochromatic K_4.

Assume that the vertices of K_{17} are labeled with the residue classes modulo 17: 0, 1, 2, ..., 16. An edge ij is colored green or red according to whether $i - j$ is a quadratic residue or a quadratic nonresidue modulo 17. The set of quadratic residues modulo 17 is the set of nonzero squares modulo 17, i.e.,

$$R_{17} = \{1, 2, 4, 9, 8, 13, 15, 16\},$$

and the set of quadratic nonresidues is the set of nonzero nonsquares, i.e.,

$$N_{17} = \{3, 5, 6, 7, 10, 11, 12, 14\}.$$

Notice that $-1 = 16 \in R_{17}$; hence $i - j$ is a quadratic residue if and only if $j - i$ is a quadratic residue. Color the edge ij green if $i - j \in R_{17}$ and red if $i - j \in N_{17}$. Suppose that there is a monochromatic K_4 on vertices a, b, c, d. We note that the coloring is translation invariant: $(i + k) - (j + k) = i - j$. Hence, we may assume that $a = 0$. Multiply each vertex by b^{-1} (the multiplicative inverse of b), and note that either no edge changes color (if $b \in R_{17}$) or every edge changes color (if $b \in N_{17}$). This is because $b^{-1}i - b^{-1}j = b^{-1}(i - j)$. In either case, we have a monochromatic subgraph K_4 on vertices 0, 1, cb^{-1}, db^{-1}. Since $1 - 0 = 1$ is a quadratic residue, the other differences, cb^{-1}, db^{-1}, $cb^{-1} - 1$, $db^{-1} - 1$, and $db^{-1} - cb^{-1}$, are all quadratic residues. By inspection of the elements of R_{17}, we find that this is impossible. This proves the lower bound $R(4, 4) > 17$ and therefore $R(4, 4) = 18$.

The above construction involving quadratic residues was discovered in 1955 by Robert E. Greenwood Jr. (1911–1995) and Andrew M. Gleason (1921–2008). Although it gives the exact Ramsey number in the case of $R(4, 4)$, the method only gives upper bounds for higher numbers. For example, using this technique we can show that $38 \leq R(5, 5)$, but other techniques show that $43 \leq R(5, 5)$. In Table 19.1, we present all the known nontrivial Ramsey numbers and bounds on some other Ramsey numbers. The notation a/b means that a and b are the best known lower and upper bounds for that particular Ramsey number. See [GRS90] and the dynamic survey by Stanislaw Radziszowski in the *Electronic Journal of Combinatorics* at http://www.combinatorics.org.

How difficult would it be to calculate $R(5, 5)$? We have the upper bound $R(5, 5) \leq R(4, 5) + R(5, 4) = 50$, but this still leaves us with an enormous computation problem in evaluating $R(5, 5)$. A naive approach, examining all $2^{\binom{49}{2}}$ labeled graphs on 49 vertices, is beyond our current computer capability.

When we consider more than two colors, the only known nontrivial Ramsey number is $R(3, 3, 3) = 17$.

a \ b	3	4	5	6	7	8	9
3	6	9	14	18	23	28	36
4		18	25	35/41	49/61	56/84	73/115
5			43/49	58/87	80/143	101/216	125/316
6				102/165	113/298	127/495	169/780
7					205/540	216/1031	233/1713
8						282/1870	317/3583
9							565/6588

TABLE 19.1: Ramsey numbers $R(a, b)$.

We now state an upper bound for 2-color Ramsey numbers (see Exercises).

Theorem 19.6. For all $a, b \geq 2$, we have $R(a, b) \leq \binom{a+b-2}{a-1}$.

It follows that

$$
\begin{aligned}
R(a, a) \; &\leq \; \binom{2a-2}{a-1} \\
&< \; 2^{2a-2} \\
&= \; 4^{a-1} \\
&< \; 4^a.
\end{aligned}
$$

For a lower bound for the diagonal Ramsey numbers $R(a, a)$, we give two non-constructive proofs. (See Chapter 12 and [ASE92].)

Theorem 19.7. If $\binom{n}{a} 2^{1-\binom{a}{2}} < 1$, then $n < R(a, a)$.

Proof 1 (Cardinality). There are $2^{\binom{n}{2}}$ green–red colorings of the $\binom{n}{2}$ edges of K_n. The number of green–red colorings of K_n with a monochromatic subgraph K_a is $|\bigcup A_S|$, where A_S is the collection of green–red colorings in which the subgraph S is monochromatic, and S ranges over all possible subgraphs of K_n isomorphic to K_a. We bound $|\bigcup A_S|$ as follows:

$$
\left| \bigcup A_S \right| \leq \sum_S |A_S| \quad \text{(Bonferroni inequalities)}
$$

$$
= 2 \cdot \binom{n}{a} 2^{\binom{n}{2} - \binom{a}{2}}
$$

$$
< 2^{\binom{n}{2}}.
$$

The equality holds because there are $\binom{a}{2}$ subgraphs K_a in K_n. Since each K_a is monochromatic, there are two choices for the color of its edges. The remaining $\binom{n}{2} - \binom{a}{2}$ edges of K_n are colored green or red arbitrarily.

As $|\bigcup A_S|$ is less than the total number of green–red colorings of K_n, we conclude that there *exists* a coloring without a monochromatic K_a. Therefore $R(a, a) > n$.   ∎

Proof 2 (Probability). Suppose that the edges of K_n are randomly and independently colored green or red. Each edge has an equal chance of being colored green or red. For each subgraph S of K_n isomorphic to K_a, let A_S be the event that S is monochromatic. Then

$$
\begin{aligned}
\Pr(A_S) &= \Pr(S \text{ is green}) + \Pr(S \text{ is red}) \\
&= 2^{-\binom{a}{2}} + 2^{-\binom{a}{2}} \\
&= 2^{1-\binom{a}{2}},
\end{aligned}
$$

and it follows that

$$
\begin{aligned}
\Pr\left(\bigcup A_S\right) &\leq \sum_S \Pr(A_S) \quad \text{(Bonferroni inequalities)} \\
&= \binom{n}{a} 2^{1-\binom{a}{2}} \\
&< 1.
\end{aligned}
$$

Since the complement of $\bigcup A_S$ occurs with positive probability, there *exists* a 2-coloring of K_n with no monochromatic K_a. Therefore $R(a, a) > n$.   ∎

Theorem 19.7 gives a lower bound for $R(a, a)$ (see Exercises):

$$
R(a, a) > 2^{a/2}, \quad a \geq 3.
$$

Thus, we conclude that

$$
\sqrt{2}^{\,a} < R(a, a) < 4^a, \quad a \geq 3.
$$

The precise growth rate of the diagonal Ramsey numbers $R(a, a)$ is unknown.

In 1995 Jeong Han Kim showed that the order of magnitude of $R(n, 3)$ is $n^2 / \log n$ (see [CG99]). The growth rate of $R(n, k)$ for arbitrary fixed k is unknown.

Exercises

1. How many nonisomorphic graphs of order 4 are there?

2. Find a self-complementary graph of order 5.

$\star$3. Show that there exists a self-complementary graph of order n if and only if n is of the form $4k$ or $k+1$.

4. Using the pigeonhole principle, show that some positive integral power of 13 ends in 0001 (base 10).

5. Let q be an odd integer greater than 1. Show that there is a positive integer n such that $2^n - 1$ is a multiple of q.

6. Show that if S is a subset of $\{1, \ldots, 2n\}$ and $|S| > n$, then there exist $x, y \in S$ with x and y relatively prime.

7. Show that if S is a subset of $\{1, \ldots, 2n\}$ and $|S| > n$, then there exist $x, y \in S$ with x a divisor of y.

8. Suppose that an $n \times n$ binary matrix contains a 1 in every row, column, and diagonal (diagonals of every length are considered here). What is the minimum number of 1s in the matrix?

†9. Show that, up to isomorphism, $K_{n,n}$ is the only triangle-free graph with $2n$ vertices and n^2 edges.

†$\star$10. Prove Turán's theorem.

11. Find a 2-coloring of the edges of K_5 having no monochromatic triangle.

†12. A *tournament* is a complete directed graph. (See Chapter 12.) Use Ramsey's theorem to show that for every n, there exists an $f(n)$ such that every tournament on $f(n)$ vertices contains a transitive subtournament on n vertices.

13. Show that every 2-coloring of the edges of K_6 yields two monochromatic triangles.

14. Find a 2-coloring of K_8 that proves that $R(3,4) > 8$.

15. Prove that $R(3,5) = 14$.

†16. Prove that $R(3,3,3) = 17$.

17. Show that if K_{327} is 5-colored, there exists a monochromatic K_3.

†18. Prove Theorem 19.6.

†19. Prove the lower bound for diagonal Ramsey numbers given at the end of the chapter.

◇20. Find lower and upper bounds for $R(10, 10)$.

Chapter 20

The Ramsey Game

In the Ramsey game, the second player has a drawing strategy if

$$2^{\binom{k}{2}-1} > \binom{n}{k}.$$

Many of the concepts and theorems of discrete mathematics can be converted into enjoyable and interesting games. Often, playing the games serves as a way to become more familiar with the mathematics involved, or to appreciate the mathematics from a different perspective. In some cases, we can determine which of two players should win the games and what the best strategies are.

We now define a game that has been investigated by many mathematicians, including Paul Erdős (1913–1996). We call the game *Graph Achievement*. Suppose that G is a graph without isolated vertices. The game is played by two players, called Solid and Dotted, on the "board" K_n, i.e., the complete graph on n vertices, for some n. The first player, Solid, chooses an edge of K_n and marks it with a solid line. The second player, Dotted, chooses an unmarked edge and marks it with a dotted line. Play continues in this way until someone has made a copy of G in his or her own marked lines. The first player to do this is the winner. If neither player succeeds in producing the goal graph, then the game is a draw.

Let's play an example of Graph Achievement with the graph K_3 (a triangle) on the board K_5. The moves of Solid and Dotted are shown in Figure 20.1. The moves are depicted on the graph and tabulated to the right of the graph. Solid's first move is 15, joining vertices 1 and 5 with a solid edge. Following a convention of Frank Harary (1921–2005), we jokingly call this move a "shrewd move" since it is completely arbitrary. Dotted responds with the move 12, joining edges 1 and 2 with a dotted line. Next, Solid makes the move 14, setting up a threat (indicated by a check mark ✓) of playing 45 and making a triangle. Thus, Dotted's move 45 is forced (indicated by an exclamation point !). Now, Solid makes the move 13, setting up a double-threat (of both 34 and 35). Dotted can only parry one of these threats, say, 34, so Solid wins on the fourth move with 35 (making the triangle 135).

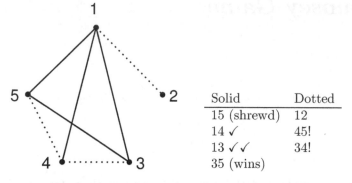

Solid	Dotted
15 (shrewd)	12
14 ✓	45!
13 ✓✓	34!
35 (wins)	

FIGURE 20.1: Solid wins a game of K_3 Achievement on K_5.

A little experimentation shows that Solid cannot force a win on the smaller board K_4. Hence, we say that the achievement number of K_3 is 5. The *achievement number* of a graph G, denoted by $a(G)$, is the smallest number n such that Solid can always win Graph Achievement for G played on K_n.

In the above discussion, we took for granted a commonplace of game theory, namely, that if there is a winner (with best possible play) of an achievement game in which players mark objects (in this case edges of a complete graph), then the winner is the first player (Solid). The rationale for this precept is that if the second player had a winning strategy, then the first player could simply appropriate it and "get there first."

Now let's consider a version of our game in which the first player to construct a K_3 *loses*. We call this game Graph Avoidance. Again, suppose that G is a graph with no isolated vertices. The game is played on a board K_n, for some n. Solid and Dotted take turns marking edges with solid lines (for Solid) and dotted lines (for Dotted). The first player to make a copy of G in his or her marked lines loses. If neither player makes a copy of G, then the game is a draw.

Now the basic principle of game theory is inverted. If there is a winner of an avoidance game, it must be the second player (Dotted).

If G is the triangle graph K_3, then Dotted can force a win in Graph Avoidance playing on K_5. (Try the game!) Since Dotted cannot force a win on K_4, we say that the avoidance number of G is 5. The *avoidance number* of a graph G, denoted by $\bar{a}(G)$, is the smallest number n such that Dotted can always win Graph Avoidance for G played on K_n.

G	K_2	P_3	$2K_2$	P_4	$K_{1,3}$	K_3	C_4	$K_{1,3}+e$	K_4-e	K_4
a	2	3	5	5	5	5	6	5	7	10
$\overline{a}$	2	3	5	5	5	5	6	5	?	?
r	2	3	5	5	6	6	6	7	10	18

TABLE 20.1: Achievement, avoidance, and Ramsey numbers.

The achievement and avoidance numbers of a triangle are related to Ramsey numbers. Recall that the Ramsey number $R(3,3) = 6$, since 6 is the smallest value of n such that if the edges of K_n are colored with two colors, there must be a monochromatic subgraph K_3. Thinking of the two colors as solid lines and dotted lines, the Ramsey number result implies that there will always be a winner of Graph Achievement for K_3 on K_6, and the same is true for Graph Avoidance.

For any graph G, we define the *graph Ramsey number* $r(G)$ to be the smallest integer n such that if the edges of K_n are colored using two colors, then there exists a monochromatic copy of G. If G is a complete graph K_m, then $r(G)$ is the same as $R(m,m)$. If G is a graph on n vertices, then $r(G) \leq R(n,n)$, since a monochromatic K_n contains a monochromatic copy of G. Obviously, $a(G) \leq r(G)$ and $\overline{a}(G) \leq r(G)$.

Table 20.1 shows achievement numbers, avoidance numbers, and graph Ramsey numbers of graphs with 2, 3, and 4 vertices. Explanations are in order for the unfamiliar graph names. The graph $2K_2$ consists of two disjoint copies of K_2. The graph $K_{1,3}+e$ consists of the graph $K_{1,3}$ with one additional edge. The graph $K_4 - e$ (sometimes jokingly called the "random graph") is the graph K_4 with one edge deleted. Verification of some of these values is called for in the exercises. The question marks indicate unknown values.

Paul Erdős and John Selfridge proved the following result about Graph Achievement for K_k played on the graph K_n: the game is a draw if $2^l > \binom{n}{k}$, where $l = \binom{k}{2} - 1$. We will prove this result shortly.

The Graph Achievement game for a complete graph is also called the *Ramsey game*. Let $n \geq k \geq 1$. Suppose that Solid and Dotted take turns marking unmarked edges of K_n. The winner is the first person to complete a K_k in his or her own marked edges. If neither player succeeds, then the game is a draw. By our basic principle of game theory, if there is a winner it must be the first player. However, it isn't clear how large n must be relative to k to ensure a win, or how the players should play this game. In fact, in the case where the game is a first player win, the winning strategy isn't known in general.

We will invoke a seminal result of Erdős and Selfridge. We define a game as follows. Let $\{A_k\}$ be a finite family of finite sets. Two players take turns choosing elements of $\bigcup_k A_k$. The first player who has chosen all the elements of some A_k is the winner. If neither player succeeds, then the game is a draw. From our basic principle, we know that only the first player can win (with best play). But how many sets are needed to guarantee that the first player can win? If the number of sets is small enough, then the second player always has a blocking strategy.

Theorem 20.1 (Erdős–Selfridge, 1973). Let $f(n)$ be the least integer for which there exist $f(n)$ sets A_k, with $|A_k| = n$ for $1 \leq k \leq f(n)$, such that the first player has a winning strategy in the game above. Then $f(n) = 2^{n-1}$.

Proof. We begin by showing that there exist 2^{n-1} sets of size n for which the first player has a winning strategy. The winning sets are the subsets of the collection $\{w, x_1, y_1, x_2, y_2, \ldots, x_{n-1}, y_{n-1}\}$ that contain w and exactly one of x_i and y_i, for $1 \leq i \leq n - 1$. Thus, there are 2^{n-1} winning subsets, each of size n. The first player has a winning strategy: on the first turn, choose w, and thereafter choose x_i if the second player chooses y_i, and vice versa. In this way, the first player chooses a winning set.

Now suppose that there are fewer than 2^{n-1} winning sets. We must show that the second player has a drawing strategy. At any stage in the game when it is the second player's turn to play, we define the "danger" of the position to indicate how close the second player is to losing. Sets from which the second player has already selected some elements represent no danger to the second player (since they cannot be completed by the first player), so we ignore such sets. We say that the danger of any other set is 2^k, where the first player has chosen k of the elements. The danger of the position is the total danger of all the sets. If the first player were to occupy all n elements of some set, then the danger of that set would be 2^n. However, we will describe a strategy by which the second player can keep the danger of the position less than 2^n, so that *a fortiori* the first player cannot win. Define the "score" of each unchosen element in the collection to be the total danger of all the sets that contain that element. The second player chooses an element with maximum score. Then the first player gets a turn. We claim that the result of the second player's turn and the first player's turn cannot be to increase the danger of the position. The second player's move removes from consideration all sets that contain the chosen element, and the first player's move doubles the score of each element in the collection of sets containing the chosen element. The second player has chosen a point with maximal score, so the result of these two moves is to decrease the danger or leave it unchanged. Since the danger of the position after the first player's first move is less than $2^{n-1} \cdot 2^1 = 2^n$, and the danger never increases, the danger never reaches 2^n. Therefore, the first player cannot complete a set. ∎

In the Ramsey game, there are $\binom{n}{k}$ winning sets each of cardinality $\binom{k}{2}$.

Corollary 20.2. The second player has a drawing strategy in the Ramsey game if

$$2^{\binom{k}{2}-1} > \binom{n}{k}.$$

The drawing strategy for the second player is given explicitly in the proof of the Erdős–Selfridge theorem.

Exercises

1. Prove that for the quadrilateral C_4, the achievement number is 6.

2. Show that Dotted wins the Graph Avoidance game for K_3, playing on K_5 but not on K_4, implying that $\overline{a}(G) = 5$.

3. Investigate a three-person version of Graph Achievement for K_3. What is the achievement number?

†4. Show that

   (a) $a(C_m) = m$, $m \geq 8$;

   (b) $a(mK_2) = 2m + 1$, $m \geq 2$.

⋆5. Define Bipartite Graph Achievement and Avoidance games as follows. Suppose that G is a bipartite graph with no isolated vertices. Let two players, Solid and Dotted, take turns marking the edges (solid or dotted, respectively) of a complete bipartite graph $K_{n,n}$. In Bipartite Graph Achievement, the first player to make a copy of G in his or her marked lines is the winner. In Bipartite Graph Avoidance, the first person to do this is the loser. Define the bipartite graph achievement number, $ba(G)$, to be the smallest n such that Solid wins Bipartite Graph Achievement on $K_{n,n}$. Define the bipartite graph avoidance number, $\overline{ba}(G)$, to be the smallest n such that Dotted wins Bipartite Graph Avoidance on $K_{n,n}$. Define the *bipartite graph Ramsey number* $br(G)$ to be the minimum n such that no matter how the edges of $K_{n,n}$ are colored with two colors, there exists a monochromatic copy of G.

(a) Illustrate a game of Bipartite Graph Achievement for C_4 on $K_{4,4}$. Show that Solid can always win.

(b) Verify the values in the table below. (The Y graph is $K_{3,1}$ with an additional edge appended to one of the vertices of degree 1.)

G	K_2	P_3	$2K_2$	P_4	$K_{1,3}$	$P_3 \cup K_2$
ba	1	2	3	3	4	3
$b\bar{a}$	1	3	3	3	4	3
br	1	3	3	3	5	3

G	C_4	P_5	$K_{2,3} - e$	$K_{1,4}$	Y	$K_{2,3}$
ba	4	4	4	6	5	6
$b\bar{a}$	4	4	4	?	4	?
br	5	4	5	7	5	9

(c) Verify the following formulas (for $m \geq 1$):

(1) $ba(K_{1,m}) = 2m - 2$;

(2) $ba(mK_2) = m + \lfloor \sqrt{m-1} \rfloor$;

(3) $ba(P_m) = \begin{cases} m - 1, & m = 2, 3, 4, \\ \lfloor (m+3)/2 \rfloor, & m \geq 5; \end{cases}$

(4) $ba(C_{2m}) = 2m$.

(d) What is the greatest value of n that you can find for which the second player can draw playing Bipartite Graph Achievement with the graph $K_{10,10}$ on the board $K_{n,n}$?

(e) Find $b\bar{a}(K_{1,4})$ and $b\bar{a}(K_{2,3})$.

6. Define Positive Triangle Achievement as follows. Two players take turns marking the edges of the complete graph K_n, for some n, using $+$ and $-$ marks. The players may choose either mark; for this reason the game is called a *choice game*. In Positive Triangle Achievement, the first player to complete a triangle with an even number of $-$ signs (a "positive triangle") wins. In this game, the goal triangle can contain marks made by both players.

(a) Show that the first player wins Positive Triangle Achievement if $n \equiv 2$ or $3 \pmod 4$, while the second player wins if $n \equiv 0$ or $1 \pmod 4$.

(b) How would you define Negative Triangle Achievement? Who wins this game on K_n?

(c) Define Positive Triangle Avoidance and Negative Triangle Avoidance. Investigate these games and demonstrate who wins on K_n for various values of n.

◇7. Let $k = 10$. Find the largest value of n that you can for which the second player has a drawing strategy in the corresponding Ramsey game.

†8. Given k, determine n such that the second player has a drawing strategy in the Ramsey game. Find the greatest value of n (relative to k) that you can.

†9. Find $\bar{a}(K_4 - e)$ and $\bar{a}(K_4)$.

†10. In 2009 Roland Bacher and Shalom Eliahou proved that every 14×15 binary matrix contains four equal entries (all 0s or all 1s) at the vertices of a square with horizontal and vertical sides. Play a two-person game based on this result. Can you find a good strategy for the first player to employ?

Chapter 21

Tic-Tac-Toe and Animal Games

Tic-tac-toe is a draw game on a 5-dimensional board of side 22.

There are 12 winning animals and 12 minimal non-winning animals (if Snaky is a winner).

In this chapter, we will investigate higher-dimensional tic-tac-toe and generalizations of tic-tac-toe known as animal achievement and avoidance games.

Tic-tac-toe is played on a 3×3 board, as shown (in two versions) in Figure 21.1. Two players, Oh and Ex, take turns placing their symbols (O and X) in unoccupied cells of the board. The first player to complete three cells in a row (horizontally, vertically, or diagonally) in his or her own symbol wins. It is well known that with best play this game is a draw (neither player can force three-in-a-row).

But what if we play this game on a larger board, say a 5×5 board? In this game, the players try to complete five cells in a row. Can one of the players force a win? Recall from Chapter 20 that in a game such as this, where both players are taking unclaimed objects (the cells), if there is a winner with best-possible play, the winner must be the first player. The

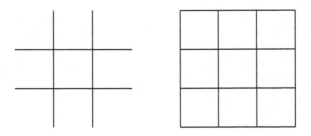

FIGURE 21.1: The tic-tac-toe board.

reason is that if the second player had a winning strategy then the first player would simply appropriate it and make the winning plays sooner.

In the tic-tac-toe game on an n-dimensional board of side k, there are

$$\frac{1}{2}[(k+2)^n - k^n]$$

winning lines (rows, columns, and diagonals). The proof of this is neat. Let T be the set $\{1, 2, 3, \ldots, k\}$. Then T^n, which consists of all strings of length n from T, is the tic-tac-toe board in n-dimensional space of side k. Let S be the set $\{0, 1, 2, 3, \ldots, k, \infty\}$. The set S^n consists of all strings of length n from the set S. Thus, S^n is a "superboard" of the tic-tac-toe board T^n. The "border" of S^n is the set of elements of S^n that contain at least one coordinate 0 or ∞. The idea is to start at a border cell and go "across" the interior cells T^n. For example, the cell $(0, 3, 1, \infty)$ gives rise to the line $(1, 3, 1, 3)$, $(2, 3, 1, 2)$, $(3, 3, 1, 1)$. The cell $(\infty, 1, 3, 0)$ yields the same line. Since there are $(k+2)^n$ elements of S^n, and we subtract the k^n elements of T^n, the number of pairs of border cells is $\frac{1}{2}[(k+2)^n - k^n]$, and this is the number of winning lines.

For example, in ordinary 3×3 tic-tac-toe, the number of winning lines is $((3+2)^2 - 3^2)/2 = 8$. These lines can be viewed as lines that join border cells in a 5×5 array. There are 16 border cells joined in 8 pairs.

It follows from the Erdős–Selfridge theorem (Theorem 20.1) that if

$$\frac{1}{2}[(k+2)^n - k^n] < 2^{k-1},$$

then the second player has a drawing strategy in tic-tac-toe on an n-dimensional board of side k. For example, the second player can draw 5-dimensional tic-tac-toe on a board of side 22.

We next discuss generalizations of 2-dimensional tic-tac-toe.

Frank Harary (1921–2005) created numerous games based on mathematical concepts and theorems that turn out to be fun and interesting to play. Professor Harary was a pioneer of graph theory; he was interested not only in the foundations of the subject, but also in applying the theory to real-life situations, including psychology and political science. Using graph theory as the basis for games is a good way to make the theory accessible.

An *animal* is a collection of edge-wise adjacent unit squares in the plane. This type of figure is also called a polyomino. The most primitive animal is Elam, consisting of a single square. From Elam grows, as in cell reproduction, the unique 2-cell animal called Domino. Domino has two children, Tic and El. Tic and El have as progeny the five 4-cell animals Fatty, Elly, Tippy, Skinny, and Knobby. From these animals are born the 5-cell animals, which generate the 6-cell animals, etc. The family tree of the animals is shown in Figure 21.2.

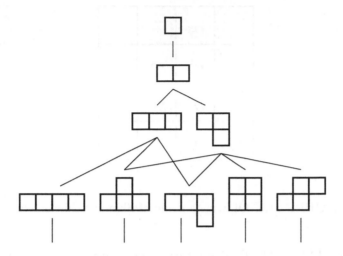

FIGURE 21.2: Evolution of the larger from the smaller animals.

No one knows a formula for the number of n-cell animals.

Any of the animals may be used as the goal configuration in an achievement game. By definition, in an achievement game the object is to be the first player to complete a certain goal. Let two players, Oh and Ex, play an achievement game on a square board decomposed into smaller squares. Oh writes an O into any small square of the board and Ex writes an X into any unoccupied square, and so on. The first player who completes in his or her symbol a configuration shaped like the predetermined goal animal is the winner. If neither player is able to do this, then the game is a draw. For each animal, an avoidance game can also be played.

We assume that Ms. Oh and Mr. Ex always make the best possible moves. This assumption of *rational play* is necessary to see exciting games. More specifically, *rational play* means that a player who has a winning position plays so that he wins or still has a winning position; if he has a drawing position, he plays so that he still has a draw; and if he has a losing position, he plays so that the loss is delayed as long as possible.

Even under the conditions of rational play, some games are more exciting than others. The game of Elam Achievement is dull. Ms. Oh, on her first turn, writes an O anywhere on the playing board and wins instantly. She has constructed in her symbol a configuration in the same shape as the

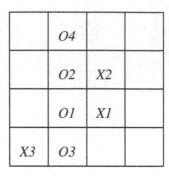

FIGURE 21.3: Oh wins a game of Skinny Achievement.

1-cell animal Elam. Mr. Ex doesn't even get a turn. Such a game is called a *trivial game* or, whimsically, a *banker's game* because of the risk involved to the eventual winner.

Oh can achieve Elam on a playing board consisting of a single square. Hence, we say that the *board number* of Elam is 1. In general, the board number of an animal is the smallest size square board on which a player can force a win.

Oh achieves Elam in a single move on a 1×1 board. Thus we say that the *move number* of Elam is 1. The move number of an animal is the number of turns necessary to achieve the animal on the smallest possible board. Animals for which the move number is equal to the number of cells in the animal are called *economical*.

As we have already indicated, in a rational two-person achievement game in which there is a winner, that winner will be the first player. In an avoidance game, the winner will be the second player.

Figure 21.3 shows an Example of Skinny Achievement on a 4×4 board in which Oh wins in six moves. This particular game does not exhibit rational play by Ex.

The biggest surprise in animal achievement theory is that there are animals which Ex can block Oh from making, even on an infinite board. The simplest example of such a non-winner animal is Fatty. Andreas Blass conceived of the idea of a domino blocking pattern. Consider the domino tiling pattern of the infinite board indicated in Figure 21.4. Any Fatty on this board must contain a complete domino. Hence Ex's strategy for blocking Fatty is simply to take the other half of any domino entered by Oh. Then Oh can never complete a domino, and thus never achieve Fatty.

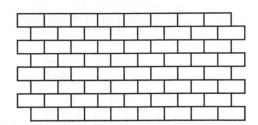

FIGURE 21.4: A domino blocking pattern for Fatty.

There are 11 known winning animals, pictured in Figure 21.5. For all other animals (except one), there is a blocking pattern which Ex can use to prevent Oh from winning the achievement game. There are 12 known minimal non-winners. The exceptional animal is Snaky (Figure 21.6), which does not have a domino blocking pattern, yet which still may be a non-winner. If Snaky is a winner, then there are 12 winners and 12 minimal non-winners. The status of Snaky has remained an open question for over 30 years.

What progress has been made on this long-standing problem of Snaky Achievement? In 1982 the author showed that there is no domino blocking pattern for Snaky. However, the lack of a domino blocking pattern doesn't mean that Snaky is a winner. It may be that Ex can draw without using a domino blocking pattern. Figure 21.7 displays a game in which Oh wins. But this is not a proof either as Ex might play better. In 2004 Nándor Sieben proved that the polyomino version of Snaky Achievement is a first player win on a 41-dimensional board.

A generalization of animal achievement and avoidance is afforded by so-called Picasso animals. A *Picasso animal* is a collection of cells on a square grid (not necessarily connected) with the property that when certain columns and rows of empty cells are eliminated, there results an animal. Figure 21.8 shows a Picasso Snaky.

Achievement and avoidance games can be played with Picasso animals. Some Picasso animals are equivalent to bipartite graphs. The equivalence is effected by turning the squares of the Picasso animals into edges of the graph. Two edges are incident at a vertex if the two cells of the Picasso animal are in the same row or column. For example, Picasso Tippy is

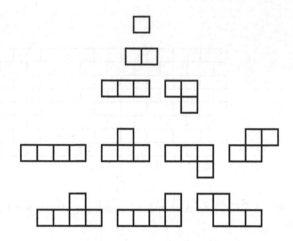

FIGURE 21.5: Animal winners.

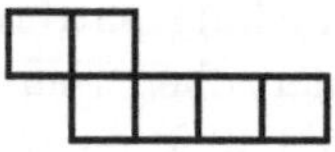

FIGURE 21.6: Snaky (is it a winner?).

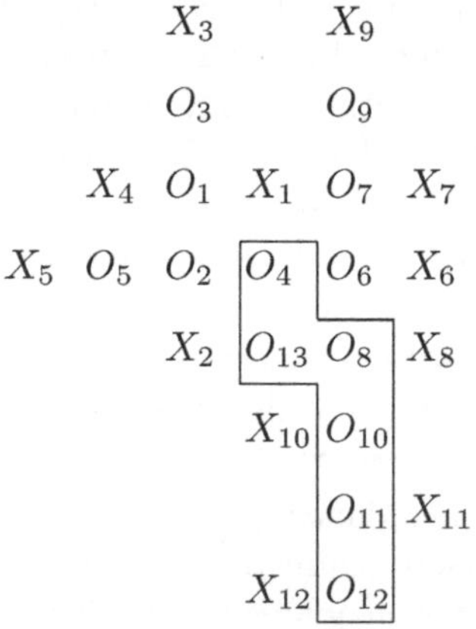

$$
\begin{array}{ccccc}
 & X_3 & & X_9 & \\
 & O_3 & & O_9 & \\
 X_4 & O_1 & X_1 & O_7 & X_7 \\
 X_5 & O_5 & O_2 & \boxed{O_4} & O_6 & X_6 \\
 & X_2 & O_{13} & O_8 & X_8 \\
 & & X_{10} & O_{10} & \\
 & & & O_{11} & X_{11} \\
 & & X_{12} & O_{12} &
\end{array}
$$

FIGURE 21.7: Oh wins a game of Snaky Achievement.

FIGURE 21.8: A Picasso Snaky.

FIGURE 21.9: A Picasso Tippy and an equivalent bipartite graph.

equivalent to the path of length 4, as shown in Figure 21.9. A complete bipartite graph $K_{m,n}$ is equivalent to the Picasso Animal based on an $m \times n$ "block." For example, Picasso Fatty is equivalent to $K_{2,2}$. For the Picasso animals that are not equivalent to bipartite graphs, the achievement and avoidance games are genuinely new games that can be fun to play.

Exercises

1. Prove that tic-tac-toe is a draw on a 5 × 5 board.

2. Find the greatest side length that you can for which the second player can draw 7-dimensional tic-tac-toe on a board of this side length.

3. In Figure 21.2, identify Elam, Domino, Tic, El, Skinny, Knobby, Elly, Fatty, and Tippy.

4. Show that Tippy Achievement is a first player winner on a 3×3 board.

5. Which animals are economical winners?

6. Draw pictures of the 12 (known) minimal non-winning animals and find a blocking strategy for each.

7. Find the board number of Picasso Fatty Achievement.

8. What is the outcome of Domino Avoidance on an $n \times n$ board with best-possible play?

9. What is the outcome of Domino Avoidance on an $n \times 1$ board with best-possible play?

10. Consider the game of avoiding Tic on a 4×4 board or a 5×5 board. Decide whether Tic is a loser or a draw on these boards.

11. Which 4-cell animal avoidance game, when played on a 3×3 board, is a first player loss?

12. Which 4-cell animals are equivalent to unique bipartite graphs?

13. Prove that every Picasso animal is a first player winner in the achievement game.

14. We know that Tippy is a winner on a 3×3 board. Is it a winner on a 5×5 board with the center square removed? What if "wrap-around" is allowed?

15. Investigate achievement and avoidance games for 3-D animals made of cubes. Give an example of a non-winning animal.

16. Animal achievement and avoidance can be played as one-color games, in which both players put the same symbol down on the board. The first player to make a copy of a predetermined animal is the winner (in the achievement game) or the loser (in the avoidance game). Who wins one-color Fatty Achievement on a 3×3 board?

17. Let Oh and Ex take turns packing 4-cell animals that have not yet been packed into a 4×5 board, with the last player able to move being the winner. Show that Oh wins the achievement game by packing Tippy first. Find the unique place in the board where Tippy must be placed in order that Oh wins.

Part VIII

Algorithms

Chapter 22

Counters

The three-digit numbers 000 through 999 (base 10) can be arranged in a circular list so that consecutive numbers differ in exactly one digit.

In this chapter, we describe some basic counting algorithms, including Gray codes. Gray codes are a popular topic in computer science because of their connections with algorithms, graphs, and puzzles like the Tower of Hanoi.

We all know how to count, but how do we tell a computer how to count? Consider how an odometer counts from 000 to 999. The count looks like

$$000, 001, 002, 003, 004, 005, 006, 007, 008, 009, 010, \ldots.$$

At each step, the units digit is advanced until a 9 is reached, and then at the next step the units digit is set back to 0 and the tens digit is advanced. Similarly, the tens digit is advanced at every tenth step until a 9 is reached, and ten steps later the tens digit is set back to 0 and the hundreds digit is advanced.

The following algorithm formalizes this procedure. For any base k and any length of string n, the algorithm produces all k^n strings of length n over the set $\{0, 1, \ldots, k-1\}$, starting with the all 0 string and ending with the all $k-1$ string.

Counting Algorithm

Let k and n be positive integers, with $k > 1$.
Let $x_0 = 0$, $x_1 = 0$, $\ldots$, $x_{n-1} = 0$.
Output $\{x_{n-1}, \ldots, x_1, x_0\}$.
While $\{x_{n-1}, \ldots, x_1, x_0\} \neq \{k-1, \ldots, k-1, k-1\}$, do:
  Set $i \leftarrow 0$.
  While $x_i = k - 1$, do: $x_i = 0$, $i \leftarrow i + 1$.
  Set $x_i \leftarrow x_i + 1$.
  Output $\{x_{n-1}, \ldots, x_1, x_0\}$.

The output for $n = 3$, $k = 2$ is

$$\{0, 0, 0\}$$
$$\{0, 0, 1\}$$
$$\{0, 1, 0\}$$
$$\{0, 1, 1\}$$
$$\{1, 0, 0\}$$
$$\{1, 0, 1\}$$
$$\{1, 1, 0\}$$
$$\{1, 1, 1\}.$$

The task of listing all subsets of a set is similar to the task of listing all binary strings (i.e., the case $k = 2$), but we only need to keep track of the selected elements, not the non-selected elements. The following algorithm lists all subsets of the set $\{1, 2, \ldots, n\}$.

Subsets Listing Algorithm

```
Let n be a positive integer.
Output {}.
Let x = {1}.
While x ≠ {n}, do:
  If the last element of x is n,
  then delete the last element of x
  and increase the new last element of x by 1;
  otherwise, append to x
  the last element of x incremented by 1.
  Output x.
```

The output for $n = 3$ is

$$\{\}, \{1\}, \{1, 2\}, \{1, 2, 3\}, \{1, 3\}, \{2\}, \{2, 3\}, \{3\}.$$

Notice that consecutive sets differ in size by one element, and this includes the first and last sets as a consecutive pair. The sets are listed in a different order from the one essentially given by the Counting Algorithm in the case $k = 2$.

In 1953 Frank Gray introduced the binary (base 2) Gray codes to solve a problem concerning mechanical switches. The *Gray code* is a method of counting in binary so that each binary string differs in only one bit from the previous string (including wrap-around, so that the first and last string are considered to be consecutive).

Generalizations of these sequences to any base are called *generalized Gray codes*. In a generalized Gray code, two consecutive strings differ in only one coordinate and by a numerical value of 1 (allowing "wrap-around") in that coordinate. Generalized Gray codes have some curious combinatorial properties. For instance, one of the exercises is an illustration of the discrete derivative and integral.

In the following algorithm, the generalized Gray code advances only the digit advanced by the normal count. This results in the code having the desired property that consecutive strings differ in only one coordinate and by a numerical value of 1 in that coordinate.

Generalized Gray Code Algorithm

Let k and n be positive integers, with $k > 1$.
Let $x_0 = 0$, $x_1 = 0$, $\ldots$, $x_{n-1} = 0$.
Let $y_0 = 0$, $y_1 = 0$, $\ldots$, $y_{n-1} = 0$.
Output $\{y_{n-1}, \ldots, y_1, y_0\}$.
While $\{x_{n-1}, \ldots, x_1, x_0\} \neq \{k-1, \ldots, k-1, k-1\}$, do:
  Set $i \leftarrow 0$.
  While $x_i = k-1$, do: $\quad x_i = 0$, $i \leftarrow i+1$.
  Set $x_i \leftarrow x_i + 1$.
  Set $y_i \leftarrow \mod(y_i + 1, k)$.
  Output $\{y_{n-1}, \ldots, y_1, y_0\}$.

Here is the output for $n = 3$ and $k = 3$.

$$\{0,0,0\}, \{0,0,1\}, \{0,0,2\},$$

$$\{0,1,2\}, \{0,1,0\}, \{0,1,1\},$$

$$\{0,2,1\}, \{0,2,2\}, \{0,2,0\},$$

$$\{1,2,0\}, \{1,2,1\}, \{1,2,2\},$$

$$\{1,0,2\}, \{1,0,0\}, \{1,0,1\},$$

$$\{1,1,1\}, \{1,1,2\}, \{1,1,0\},$$

$$\{2,1,0\}, \{2,1,1\}, \{2,1,2\},$$

$$\{2,2,2\}, \{2,2,0\}, \{2,2,1\},$$

$$\{2,0,1\}, \{2,0,2\}, \{2,0,0\}$$

Exercises

◇1. Implement the Counting Algorithm in your favorite computer language.

  2. Give the output of the Subsets Listing Algorithm for $n = 4$.

◇3. Implement the Subsets Listing Algorithm in your favorite computer language.

  4. Give the Gray code for $n = 3$ and $k = 2$.

◇5. Implement the Generalized Gray Code Algorithm in your favorite computer language.

◇6. Write a computer program to list the subsets of an n-element set that have an even number of elements.

◇7. Write a computer program to recursively output the binary Gray code of length n using the binary Gray code of length $n - 1$.

  8. A normal counter (base 10) reads 1123094. What is the corresponding Gray code count? A Gray code counter (base 10) reads 3210109. What is the corresponding normal count?

†9. Research the Tower of Hanoi puzzle and find out how it is related to the Gray code.

 10. A *Hamiltonian circuit* of a finite graph G, named after William Rowan Hamilton (1805–1865), is a circular sequence of all the vertices of G such that every two consecutive vertices in the sequence are adjacent in G. Show how a generalized Gray code is equivalent to a Hamiltonian circuit of a certain graph. What is the graph?

Chapter 23

Listing Permutations and Combinations

There is a simple way to list all the permutations of an n-element set in lexicographic order.

There is a simple way to list all the k-element combinations of an n-element set in lexicographic order.

In this chapter, we describe algorithms to list permutations and combinations of a set. Such algorithms are useful in a variety of mathematics and computer science applications. We will conclude with a wonderful "minimal change" listing of permutations known as the Johnson–Trotter algorithm.

Consider the following ordering of the 24 permutations of $\{1, 2, 3, 4\}$ (read left-to-right and top-to-bottom):

$$\begin{array}{cccccc}
1234 & 1243 & 1324 & 1342 & 1423 & 1432 \\
2134 & 2143 & 2314 & 2341 & 2413 & 2431 \\
3124 & 3142 & 3214 & 3241 & 3412 & 3421 \\
4123 & 4132 & 4213 & 4231 & 4312 & 4321.
\end{array}$$

In our notation, the permutation 1324, for example, is the one that maps $1 \to 1$, $2 \to 3$, $3 \to 2$, and $4 \to 4$.

This ordering is called the *lexicographic* or *dictionary* ordering. The permutations are listed in the order that they would appear in a dictionary if the "alphabetical" order of the numbers is 1, 2, 3, 4.

Here is an algorithm that lists the permutations of the set $\{1, \ldots, n\}$ in lexicographic order. To see how the algorithm works, try to determine what the permutation following 2431 should be (checking your answer with the listing above).

187

Permutations Listing Algorithm

Let n be a positive integer greater than 1.
Let $x_1, \ldots, x_n = 1, \ldots, n$.
Output $x_1, \ldots, x_n$.
While $x_1, \ldots, x_n \neq n, \ldots, 1$, do:
  Set $i^* \leftarrow n - 1$.
  While $x_{i^*} > x_{i^*+1}$, do: $i^* \leftarrow i^* - 1$.
  Set $j^* \leftarrow n$.
  While $x_{j^*} < x_{i^*}$, do: $j^* \leftarrow j^* - 1$.
  Interchange the values of x_{i^*} and x_{j^*}.
  Set $x_{i^*+1}, \ldots, x_n \leftarrow x_n, \ldots, x_{i^*+1}$.
  Output $x_1, \ldots, x_n$.

Notice that to find the next permutation in the sequence, we leave unaltered as many digits as possible at the left. To do this, we find the rightmost number a whose right neighbor b satisfies $a < b$. Then a is exchanged for the next greater available number, and the places to the right are filled with the available numbers in increasing order.

Now we describe an algorithm that lists all k-element combinations of the set $\{1, \ldots, n\}$ in lexicographic order. To find the next subset in the order, we want to simply add 1 to the last element in the current subset. When this isn't possible (i.e., the last element is n), then we drop this last element and, if possible, increment the new last element and include its successor. This isn't possible when the new last element is $n - 1$; in this case, we need to go further back in the list to find an element to increment. The following algorithm makes this procedure precise.

Combinations Listing Algorithm

Let n and k be positive integers with $1 \leq k \leq n$.
Let $x = \{1, \ldots, k\}$.
Output x.
While $x \neq \{n - k + 1, \ldots, n\}$, do:
  Set $i^* \leftarrow k$.
  While $x_{i^*} = n - k + i^*$, do: $i^* \leftarrow i^* - 1$.
  Set $x_{i^*} \leftarrow x_{i^*} + 1$.
  Set $x_{i^*+1}, \ldots, x_k \leftarrow x_{i^*} + 1, \ldots, x_{i^*} + k - i^*$.
  Output x.

Here is the output for $n = 6$ and $k = 3$.

$$\{1,2,3\}, \{1,2,4\}, \{1,2,5\}, \{1,2,6\}, \{1,3,4\},$$

$$\{1,3,5\}, \{1,3,6\}, \{1,4,5\}, \{1,4,6\}, \{1,5,6\},$$

$$\{2,3,4\}, \{2,3,5\}, \{2,3,6\}, \{2,4,5\}, \{2,4,6\},$$

$$\{2,5,6\}, \{3,4,5\}, \{3,4,6\}, \{3,5,6\}, \{4,5,6\}$$

For our final topic of this chapter, let's return to permutations. We have seen that we can list the permutations of $\{1, 2, \ldots, n\}$ in lexicographic order. Can we also list permutations in such a way that successive permutations differ only by a transposition of elements? Can you find a way to list the permutations of $\{1, 2, 3\}$ so that successive permutations differ by a transposition of adjacent elements? Such a listing is afforded by the Johnson–Trotter algorithm, discovered in 1962-63 by Selmer M. Johnson and Hale F. Trotter. To each integer we assign a direction, left or right. Ultimately, we don't care about the directions, but they are used in the algorithm. We represent, say, a 3 with a "left" direction as

$$\overleftarrow{3} \, .$$

Initially, all integers are directed to the left. We say that an integer is *mobile* if it is larger that its immediate neighbor (if one exists) in its given direction. Each step of the algorithm consists of transposing the largest mobile integer with its neighbor in its given direction, and changing the direction of all larger integers. This transpires until there are no mobile integers.

Johnson–Trotter Algorithm

```
Let n be a positive integer.
Assign to each integer 1, ..., n a ''left'' direction.
Output {1, 2, ..., n}.
While there exists a mobile integer, do:
  Let l be the largest mobile integer.
  Transpose l and its neighbor in the direction of l.
  Reverse the direction of all integers larger than l.
  Output the current permutation.
```

Here is the output for $n = 3$.

$$\overset{\leftarrow}{1}, \overset{\leftarrow}{2}, \overset{\leftarrow}{3}$$

$$\overset{\leftarrow}{1}, \overset{\leftarrow}{3}, \overset{\leftarrow}{2}$$

$$\overset{\leftarrow}{3}, \overset{\leftarrow}{1}, \overset{\leftarrow}{2}$$

$$\overset{\rightarrow}{3}, \overset{\leftarrow}{2}, \overset{\leftarrow}{1}$$

$$\overset{\leftarrow}{2}, \overset{\rightarrow}{3}, \overset{\leftarrow}{1}$$

$$\overset{\leftarrow}{2}, \overset{\leftarrow}{1}, \overset{\rightarrow}{3}$$

Notice that the list is circular, that is, the first and last permutations differ by a transposition of adjacent numbers.

Exercises

◇1. Write a program to compute $n!$ for any nonnegative integer n.

◇2. Write a program to compute binomial coefficients $\binom{n}{k}$.

3. An unusual dictionary contains as "words" all permutations of the 26 letters of the alphabet. The words are listed in alphabetical order (in all capital letters), starting with

ABCDEFGHIJKLMNOPQRSTUVWXYZ

and ending with

ZYXWVUTSRQPONMLKJIHGFEDCBA.

What word in the dictionary comes immediately after

JMZORTXLBPSYWVINGDUQEKHFCA?

◇4. Implement the Permutations Listing Algorithm.

◇5. Implement the Combinations Listing Algorithm.

6. Professor Bumble's favorite permutations of integers are those that map even integers to odd integers and odd integers to even integers. Of the permutations of the set $\{1, 2, \ldots, 2n\}$, where $n \geq 1$, how many are among Professor Bumble's favorites? Describe a procedure that lists these permutations.

†7. Explain why the Johnson–Trotter algorithm works. Do you see a resemblance between the lists produced and Gray codes?

8. In the Johnson–Trotter algorithm, how many steps occur between the permutations $\{1, 2, \ldots, n\}$ and $\{n, n-1, \ldots, 1\}$? What directions do the integers have for the permutation $\{n, n-1, \ldots, 1\}$?

Chapter 24

Sudoku Solving and Polycube Packing

There is a simple algorithm that solves Sudoku puzzles instantly.

There are 128 ways to pack the seven 4-cube polycubes into a $2 \times 2 \times 7$ box.

In this chapter, we will discuss a multi-purpose tool called an Exact Cover Algorithm. We show how to use this algorithm to solve Sudoku puzzles and polycube packing problems. The algorithm furnishes all solutions to these problems.

An *exact cover problem* consists of a finite S and a collection of subsets of S. The goal is to find an *exact cover* of S, that is, a subcollection of the subsets whose disjoint union is S. For example, suppose that

$$S = \{1, 2, 3, 4, 5\}$$

and the subsets are $X_1 = \{1\}$, $X_2 = \{2\}$, $X_3 = \{3, 4, 5\}$, and $X_4 = \{2, 3, 4, 5\}$. Then there are two exact covers: $\{X_1, X_2, X_3\}$ and $\{X_1, X_4\}$.

An exact cover problem can be given in terms of a binary matrix (where each entry is 0 or 1). The columns of the matrix represent the elements of the base set and the rows represent the subsets. Put a 1 in a given position if the corresponding element is in the corresponding subset; put a 0 in the position otherwise. The binary matrix associated with our example is

$$\begin{bmatrix} 1 & 0 & 0 & 0 & 0 \\ 0 & 1 & 0 & 0 & 0 \\ 0 & 0 & 1 & 1 & 1 \\ 0 & 1 & 1 & 1 & 1 \end{bmatrix}.$$

The columns represent the elements 1, 2, 3, 4, and 5, and the rows represent the subsets X_1, X_2, X_3, and X_4. The exact cover problem is to find a set of rows that have exactly one 1 in each column. The two solutions are (1) the first three rows and (2) the first and fourth rows.

An Exact Cover Algorithm is a procedure for solving an exact cover problem. Here is a straightforward version proposed by Donald Knuth. No faster general Exact Cover Algorithm is known.

Exact Cover Algorithm

```
Let M be a binary matrix.
If M is empty, then the problem is solved.
Otherwise, choose a column c with the minimum number of 1s.
If this number is zero, then terminate unsuccessfully.
Choose, in turn, all rows r such that M[r, c] = 1.
Include r in a partial solution.
For each j such that M[r, j] = 1, do:
  Delete column j from M.
  For each i such that M[i, j] = 1, delete row i from M.
Repeat this algorithm recursively on the reduced matrix M.
```

The direction to choose column c with the minimum number of 1s is arbitrary; it has been found empirically that this criterion results in fast run-times. Once c is chosen, we should look at each row in turn that "covers" the element that column c represents. Each such row, included in a partial solution, may also cover other elements of the given set, and so the columns representing these elements are deleted, and any rows that duplicate the coverage of these elements are deleted. The program calls itself on the reduced matrix, essentially cloning copies of itself.

We describe how to apply our Exact Cover Algorithm to the problem of solving Sudoku puzzles and the problem of packing polycubes into a box.

In a Sudoku puzzle, the goal is to fill in the blank cells of a 9×9 grid so that every row, column, and 3×3 box contains every digit from 1 to 9. See the example puzzle in Figure 24.1.

We convert the Sudoku puzzle into an exact cover problem by forming a 729×324 binary matrix that encodes all possible ways to put a number into the Sudoku grid. As there are 9 numbers and 81 cells, there are $9 \cdot 81 = 729$ choices altogether. Thus, our binary matrix has 729 rows. Each row has exactly four 1s, corresponding to conditions that are satisfied when a number is placed in the grid. The conditions are of four types:

- xOy means that cell (x, y) is occupied;

- xRy means that number x is in row y;

- xCy means that number x is in column y;

- xBy means that number x is in block y.

			8					3
5		2		3			1	8
		6		5	7		2	
2				7		1		
	4	3				8	5	
		5		2				6
	7		2	4		6		
6	5			8		2		1
9					5			

FIGURE 24.1: A Sudoku puzzle.

These conditions are represented by the $4 \cdot 9^2 = 324$ columns.

The given numbers in a Sudoku puzzle are represented by given rows of the matrix. Solving the Sudoku puzzle is equivalent to extending the set of given rows to an exact cover of the matrix. Since an exact cover contains exactly one 1 in each column, there is exactly one i (where $1 \leq i \leq 9$) in each row, column, and block of the Sudoku grid and each cell is occupied by some number.

A computer implementation of our Exact Cover Algorithm instantly solves any Sudoku puzzle or determines that there is no solution. For instance, the puzzle in Figure 24.2 isn't solvable because there is no way to put a 4 in the top-left block. The program discovers this since there is no 1 in column $4B1$. The algorithm has a nice symmetry in that the search is over all the digits that can fit in a cell and all the places a digit can fit in a row, column, or block.

By the way, there are exactly $6670903752021072936960 \doteq 6.7 \times 10^{21}$ different Sudoku boards, as found in 2005 by Bertram Felgenhauer and Frazer Jarvis.

Now let's consider a 3-D polycube packing problem. We will refer to the figures as either polycubes or animals. As shown in Figure 24.3, there are seven animals composed of four cubes. Two animals are considered the same if one can be rotated and/or reflected to form the other.

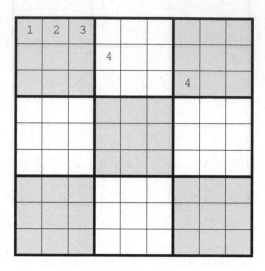

FIGURE 24.2: An unsolvable Sudoku puzzle.

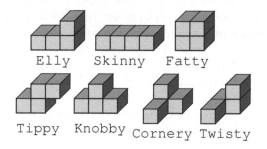

FIGURE 24.3: The seven four-cube 3-D animals.

29	30	31	32	33	34	35
22	23	24	25	26	27	28

15	16	17	18	19	20	21
8	9	10	11	12	13	14

FIGURE 24.4: A labeled $2 \times 2 \times 7$ box

We label the animals 1, 2, 3, 4, 5, 6, and 7. The particular order isn't important, but let's say that Skinny is labeled 1. We label the cells of the box 8 through 35, as shown in Figure 24.4.

The placement of an animal in the box is represented by a binary vector of length 35. The first seven coordinates of the vector tell which animal is packed. The other coordinates tell where it is packed. For example, the binary vector of length 35 consisting of all 0s except 1s in positions 1, 8, 9, 10, 11, and 12 represents a Skinny in positions 8, 9, 10, 11, and 12. Considering all possible placements of all seven animals, there are 399 such vectors. When we run our Exact Cover Algorithm on this set of vectors, we find 128 exact covers. One of these (indicating the nonzero coordinates) is

$$\{1, 9, 10, 11, 12\}, \{2, 22, 23, 29, 30\}, \{3, 25, 26, 27, 33\}, \{4, 13, 14, 19, 20\},$$

$$\{5, 8, 15, 16, 17\}, \{6, 21, 28, 34, 35\}, \{7, 18, 24, 31, 32\}.$$

In every solution, the Skinny animal is placed one cell away from a corner. Since there are eight symmetries of the box, there are $128/8 = 16$ non-equivalent packings of the eight animals.

A wealth of polycube packings is shown on the "Polycubes" page at http://www.geocities.com/alclarke0/PolyPages/Polycubes.html.

Exercises

◇1. Implement the Exact Cover Algorithm to solve Sudoku puzzles. Try your program on the puzzle of Figure 24.1.

2. In a Sudoku puzzle, what is the maximum number of given numbers that allow for more than one solution?

◇3. (a) How many solutions are there to a Sudoku puzzle with the first six rows given as

$$\{1, 2, 3, 4, 5, 6, 7, 8, 9\}$$

$$\{4, 5, 6, 7, 8, 9, 1, 2, 3\}$$

$$\{7, 8, 9, 1, 2, 3, 4, 5, 6\}$$

$$\{3, 6, 4, 9, 7, 2, 5, 1, 8\}$$

$$\{5, 1, 2, 3, 6, 8, 9, 4, 7\}$$

$$\{9, 7, 8, 5, 1, 4, 3, 6, 2\}?$$

(b) A *Latin square* is an $n \times n$ array in which each row and column contains all the integers from 1 to n. How many 4×4 Latin squares are there?

(c) Can you find a relationship between the problems in (a) and (b)?

◇4. Prove that the five 4-cell animals cannot be packed into a 4×5 board. However, two sets of these animals can be packed into a 5×8 board or a 4×10 board. How many ways can each such packing be done?

◇5. There are twelve 5-cell animals. How many ways can they be packed into

(a) a 3×20 box?

(b) a 5×12 box?

(c) a 6×10 box?

◇6. How many ways can the twelve 5-cell animals be packed into a 6×10 box so that each of them touches the border of the box?

◇7. How many ways can you pack eight dominoes (2-cell polyominoes) into a 4×4 board? How many ways can you pack 18 dominoes into a 6×6 board?

Note. These are instances of the *dimer problem*, which asks for the number of ways to pack an $m \times n$ board with dominoes (where at least one of m and n is even).

◇8. What happens in the dimer problem (see previous exercise) if two squares of the board of opposite checkerboard color are missing? What choice of the two squares results in the greatest possible number of domino tilings? What choice results in the least number?

◇9. How many ways can you pack eight dominoes into a 4×4 toroidal board (i.e., allowing wrap-around)?

◇10. Determine the number of ways that an $n \times n$ board can be packed with dominoes and/or 1-cell animals, where $1 \leq n \leq 4$.

◇11. How many ways can eight mutually hostile Queens be placed on an 8×8 chessboard so that no Queen attacks another? Solve the problem using a generalized Exact Cover Algorithm. Placing a Queen on a square "covers" the row and column containing that square, as well as the diagonal(s) containing that square. Every row and column must be covered exactly once and every diagonal may be covered at most once.

Appendix A

Hints and Solutions to Exercises

The computer code shown uses *Mathematica*.

Chapter 1

1. There are $2^4 = 16$ subsets: $\emptyset$, $\{A\}$, $\{B\}$, $\{C\}$, $\{D\}$, $\{A, B\}$, $\{A, C\}$, $\{A, D\}$, $\{B, C\}$, $\{B, D\}$, $\{C, D\}$, $\{A, B, C\}$, $\{A, B, D\}$, $\{A, C, D\}$, $\{B, C, D\}$, $\{A, B, C, D\}$.

2. By the product rule, there are $4 \times 2 \times 3 = 24$ choices.

$\diamond$3. 333

4. There are $4 \cdot 2 \cdot 10 = 80$ choices.

5. There are 10^3 choices for the first three digits, 26 for the letter, and 10 for the last digit. Hence, there are

$$10^3 \cdot 26 \cdot 10 = 260000$$

different licenses.

6. There are $3^{10} = 59049$ such strings.

7. From the collection of 2^{10} subsets of $\{a, b, c, d, e\}$, we must remove those in which both a and b appear. Since there are 2^8 of these subsets, the number of desired subsets is $2^{10} - 2^8 = 1024 - 256 = 768$.

8. There are 2^{99} binary strings of length 99. In half of these, the sum of the elements is an odd number. The reason is that the complement of a string, formed by changing each 0 to a 1 and each 1 to a 0, has a sum of elements equal to 99 minus the sum of the elements of the original string. If the original sum is odd, then the new sum is even, and *vice versa*. Therefore, the number of desired strings is $2^{99}/2 = 2^{98}$.

9. There are four choices for the image of each of the three elements in the domain of the function. Hence, there are $4^3 = 64$ functions.

10. There are n choices for the image of each of the n elements of the domain. Hence, there are n^n functions.

11. There are n choices for the image of each even number and n choices for the image of each odd number. Hence, there are $n^n \cdot n^n = n^{2n}$ functions.

12. yes

13. If the White King is in a corner, then there are $64 - 4 = 60$ places to put the Black King. If the White King is on a side, not in a corner, then there are $64 - 6 = 58$ places to put the Black King. If the White King is in the 6×6 "inner board," then there are $64 - 9 = 55$ places to put the Black King. So altogether there are

$$4 \cdot 60 + 24 \cdot 58 + 36 \cdot 55 = 3612$$

arrangements.

Chapter 2

1. The number of orderings is $8! = 40320$.

2. The number of permutations is $P(10, 6) = 10 \cdot 9 \cdot 8 \cdot 7 \cdot 6 \cdot 5 = 151200$.

$\diamond$3. 70

4. There are $3!$ different orderings of the cities in Germany, $4!$ orderings of the cities in France, and $5!$ orderings of the cities in Spain. Hence, there are $3!4!5! = 17280$ different itineraries.

5. The number of arrangements is $(3 + 4 + 5)!/(3!4!5!) = 27720$.

6. There are $4!$ ways to order the astronomy books, $5!$ ways to order the medical books, and $6!$ ways to order the religious books. The groups of books can be ordered in $3!$ ways. Hence, there are $4!5!6!3! = 12{,}441{,}600$ orderings of all the books.

7. The word RHODODENDRON has two Rs, one H, three Os, three Ds, one E, and two Ns, totaling 12 letters. Hence, the number of ways of arranging the letters is $12!/(2!1!3!3!1!2!) = 3326400$.

8. The number of one-to-one functions is $P(7, 3) = 7 \cdot 6 \cdot 5 = 210$.

9. The total number of functions is n^n and the number of one-to-one functions is $n!$. Hence, the number of functions that are not one-to-one is $n^n - n!$.

10. There are n^2 ordered pairs of elements from the given set. Each ordered pair can be included or not included in a relation. Hence, there are 2^{n^2} relations.

11. Without the restriction that class A received the wrong lecture, there are 5! possible orders of the lectures. We must subtract the number for which A received the right lecture. This is 4! (as the other four lectures can be permuted arbitrarily). Hence, the number of possible orderings is $5! - 4! = 120 - 24 = 96$.

12. Without the restriction that A and B cannot both be performed, the number of programs is $P(12,3) = 12!/9! = 1320$. We must subtract from this those programs that include both A and B. There are ten choices for the third song in such a program, and there are $3! = 6$ ways to permute the three songs. Hence, the total number of satisfactory programs is $1320 - 10 \cdot 6 = 1260$.

13. The number of selections is $C(10,3) = 10!/(3!7!) = 10 \cdot 9 \cdot 8/6 = 120$.

14. $C(66,4) = 7059052$

15. $C(20,10) = 20!/(10!10!) = 184756$

16. By the binomial theorem, the coefficient is $\binom{20}{10} = 184756$.

17. The tenth row of Pascal's triangle is 1, 10, 45, 120, 210, 252, 210, 120, 45, 10, 1. Hence

$$(a+b)^{10} = a^{10} + 10a^9 b + 45a^8 b^2 + 120a^7 b^3 + 210a^6 b^4 + 252a^5 b^5$$
$$+ 210a^4 b^6 + 120a^3 b^7 + 45a^2 b^8 + 10ab^9 + b^{10}.$$

18. $\binom{n}{2} = n!/(2!(n-2)!) = n(n-1)/2$

   $\binom{n}{3} = n!/(3!(n-3)!) = n(n-1)(n-2)/6$

19. $\binom{20}{5}$

20. $-2^7 \binom{10}{7}$

21. $\binom{20}{10}$

22. $\binom{20}{4}$

23. The formula is equivalent to the relation $k!C(n,k) = P(n,k)$.

24.
$$\binom{n-1}{k-1} + \binom{n-1}{k} = \frac{(n-1)!}{(k-1)!(n-k)!} + \frac{(n-1)!}{k!(n-k-1)!}$$

$$= \frac{(n-1)! \cdot k}{k!(n-k)!} + \frac{(n-1)! \cdot (n-k)}{k!(n-k)!}$$

$$= \frac{(n-1)! \cdot n}{k!(n-k)!}$$

$$= \frac{n!}{k!(n-k)!}$$

$$= \binom{n}{k}$$

†25. The coefficient of $a^{n-k}b^k$ is the number of ways of selecting $n-k$ factors $a+b$ that contribute as to the expansion of $(a+b)^n$ (the other factors contribute bs).

26. We are evidently looking for a solution to the equation $\binom{2n}{n} = 924$. A check of Pascal's triangle shows that the only solution to this equation is $n = 6$. So there are 12 students in the class.

27. Let L and R be the number of left steps and right steps in a path, respectively. Then $L + R = k$ and $R - L = n$, so that $R = (n+k)/2$. Hence, the number of paths is $C(k, (n+k)/2)$.

◇28.
```
c[0, 0] = 1; c[n_, 0] = 1; c[n_, n_] := 1;
c[n_, k_] := c[n, k] = c[n - 1, k] + c[n - 1, k - 1];

c[100,50]
100891344545564193334812497256
```

29. $C(n,0) + C(n,1) + C(n,2)$

30. $C(n,0) + C(n,1) + C(n,2) + C(n,3)$

Chapter 3

1. (a)
$$\binom{n}{k} = \frac{n!}{k!(n-k)!} = \frac{n(n-1)!}{k(k-1)!(n-k)!} = \frac{n}{k}\binom{n-1}{k-1}$$

(b)

$$\binom{-n}{k} = (-1)^k \binom{n+k-1}{k}$$

$$= \frac{(-n)(-n-1)(-n-2)\ldots(-n-k+1)}{k!}$$

$$= \frac{(-1)^k n(n+1)(n+2)\ldots(n+k-1)}{k!}$$

$$= (-1)^k \binom{n+k-1}{k}$$

2. We use a multiplication trick. Let

$$S = 1\cdot2 + 2\cdot3 + 3\cdot4 + \cdots + n(n+1).$$

Multiply by 3:

$$3S = 1\cdot2\cdot3 + 2\cdot3\cdot(4-1) + 3\cdot4\cdot(5-2) + \cdots + n(n+1)(n+2-(n-1))$$

$$= 1\cdot2\cdot3 + 2\cdot3\cdot4 - 1\cdot2\cdot3 + 3\cdot4\cdot5 - 2\cdot3\cdot4 + \cdots$$

$$+ n(n+1)(n+2) - (n-1)n(n+1).$$

This telescoping series collapses to yield

$$3S = n(n+1)(n+2)$$

and hence

$$S = n(n+1)(n+2)/3.$$

3. Using the same multiplication trick as in the previous exercise, we can show that

$$1\cdot2\cdot3 + 2\cdot3\cdot4 + \cdots + n\cdot(n+1)\cdot(n+2) = n(n+1)(n+2)(n+3)/4.$$

4. (a) The left side counts the ways of selecting n elements from the set $\{1,2,3,\ldots,2n-1\}$. Suppose that k elements are chosen from the subset $\{n+1,\ldots,2n-1\}$ and the remaining $n-k$ elements are chosen from the subset $\{1,2,3,\ldots,n\}$. Since $\binom{n}{n-k} = \binom{n}{k}$, the right side counts these possibilities, for $0 \le k \le n$.

(b) The left side counts the ways of selecting n elements from the set $\{1,2,3,\ldots,3n\}$. Suppose that k elements are chosen from the subset $\{1,2,3,\ldots,2n\}$ and the remaining $n-k$ elements are chosen from the subset $\{2n+1,\ldots,3n\}$. Since $\binom{n}{n-k} = \binom{n}{k}$, the right side counts these possibilities, for $0 \le k \le n$.

5. (a)

$$\binom{n}{k} = \frac{n!}{k!(n-k)!} = \frac{(n-k+1)n!}{k(k-1)!(n-k+1)!} = \frac{n-k+1}{k}\binom{n}{k-1}$$

(b) From part (a), we have $\binom{n}{k}k = (n-k+1)\binom{n}{k-1}$. Hence, $\binom{n}{k} \geq \binom{n}{k-1}$ if and only if $k \leq n-k+1$, which is equivalent to $k \leq (n+1)/2$. This proves our result.

6. From the previous problem, we have

$$\binom{n}{k} = \frac{n-k+1}{k}\binom{n}{k-1}$$

$$\binom{n}{k}\frac{n-k}{k+1} = \binom{n}{k+1},$$

and so

$$\binom{n}{k}^2\frac{n-k}{k+1} = \frac{n-k+1}{k}\binom{n}{k-1}\binom{n}{k+1}.$$

The inequality follows if we can show that

$$\frac{n-k}{k+1} \leq \frac{n-k+1}{k},$$

and this inequality is equivalent to $0 \leq n+1$.

7. Apply Pascal's identity to the binomial coefficient $\binom{m}{i}$ and collapse the resulting telescoping sum.

8. To compute the number of collisions, we may as well assume that the particles pass through each other and simply count the number of passes. In returning to its original position, each particle passes each other particle twice. Hence, the total number of particle–particle interactions is $2 \cdot 5 \cdot 4/2 = 20$.

9. In order to group $2n$ people into n pairs, we line up the $2n$ people, which may be done in $(2n)!$ ways, and take the first two as the first pair, the second two as the second pair, and so on. We have committed a lot of double-counting. We need to divide by the double-counting within pairs, $(2!)^n$, and the double-counting due to permutations of pairs, $n!$. Hence, the number of pairings is

$$\frac{(2n)!}{2^n n!} = \binom{2n}{n}\frac{n!}{2^n}.$$

10. The expression is equal to the number of ways to group 4700 people into 100 groups of size 47.

11. The binomial coefficient $\binom{m+n+1}{n+1}$ counts the number of $(n+1)$-element subsets of the set $\{1, \ldots, m+n+1\}$. The $(n+1)$st element can be any number between $n+1$ and $m+n+1$. This means that the other elements must be chosen from the set $\{1, \ldots, n+i\}$, where $0 \le i \le m$. These cases are counted by the right side of the relation.

12. The expression is equal to

$$x\frac{d}{dx}\left(x\frac{d}{dx}(1+x)^n\right)$$

evaluated at $x = 1/3$. Elementary calculus shows that this simplifies to

$$\left(\frac{4}{3}\right)^n \frac{n(n+3)}{16}.$$

13. The main idea is that in order to produce a 10 in a binary string, we need to switch from a run of 0s to a 1 and then from a run of 1s to a 0. There are four cases to consider.

Case (1): The string begins with 0 and ends with 0. There are $2k$ places where switches occur. Hence, the number of desired strings is $C(n-1, 2k)$.

Case (2): The string begins with 0 and ends with 1. There are $2k+1$ places where switches occur (the last switch being from a run of 0s to a run of 1s). Hence, the number of desired strings is $C(n-1, 2k+1)$.

Case (3): The string begins with 1 and ends with 0. There are $2k-1$ places where switches occur. Hence, the number of desired strings is $C(n-1, 2k-1)$.

Case (4): The string begins with 1 and ends with 1. There are $2k$ places where switches occur (the last switch being from a run of 0s to a run of 1s). Hence, the number of desired strings is $C(n-1, 2k)$.

Therefore, the total number of desired strings is

$$C(n-1, 2k) + C(n-1, 2k+1) + C(n-1, 2k-1) + C(n-1, 2k)$$
$$= C(n, 2k+1) + C(n, 2k) = C(n+1, 2k+1).$$

†⋆14. Obviously $S_0(n) = n$. The formula for $S_1(n)$ can be obtained by noting that $2S_1(n) = (n+1)n$ and hence $S_1(n) = n(n+1)/2$. To find $S_2(n)$ and $S_3(n)$, we use the following technique. The sum

$$\sum_{i=1}^{n}[(i+1)^{k+1} - i^{k+1}]$$

is a telescoping series. Hence

$$(n+1)^{k+1} - 1 = \sum_{i=1}^{n} [(i+1)^{k+1} - i^{k+1}]$$

$$= \sum_{i=1}^{n} \sum_{j=0}^{k} \binom{k+1}{j} i^j$$

$$= \sum_{j=0}^{k} \binom{k+1}{j} S_j(n).$$

Therefore

$$S_k(n) = \frac{1}{k+1} \left[(n+1)^{k+1} - 1 - \sum_{j=0}^{k-1} \binom{k+1}{j} S_j(n) \right].$$

We compute

$$S_2(n) = \frac{1}{3} \left[(n+1)^3 - 1 - \binom{3}{0} n - \binom{3}{1} \frac{n(n+1)}{2} \right]$$

$$= \frac{n(n+1)(2n+1)}{6}$$

and, similarly,

$$S_3(n) = \left(\frac{n(n+1)}{2} \right)^2.$$

The fact that $S_k(n)$ is a monic polynomial in n of degree $k+1$ is clear from our method.

◇15.
```
s[0] := n;
s[k_] := (1/(k + 1))((n + 1)^(k + 1) - 1 -
Sum[Binomial[k + 1, j]s[j], {j, 0, k - 1}])

s[10]
n(6n^{10}+33n^9+55n^8-66n^6+66n^4-33n^2+5)/66
```

16. $(1+x)^{-4} = 1 - 4x + 10x^2 - 20x^3 + 35x^4 - 56x^5 - 84x^6 - 120x^7 + \cdots$

◇17.
```
Series[(1 + 3x)^(-7), {x, 0, 10}]

1-21x+252x^2-2268x^3+17010x^4-112266x^5+673596x^6
-3752892x^7+19702683x^8-98513415x^9-472864392x^10+...
```

†18. In the expansion of $(x_1+x_2+\cdots+x_k)^n$, the coefficient of $x_1^{\alpha_1}x_2^{\alpha_2}\ldots x_k^{\alpha_k}$, where the α_i are nonnegative integers such that $\alpha_1+\alpha_2+\cdots+\alpha_k = n$, is the number of ways of selecting α_i factors that contribute an x_i, for each i. This is the same as the number of ways of arranging the letters of an n-element word, where there are α_i equivalent letters, for each i. These arrangements are counted by the multinomial coefficient

$$\binom{n}{\alpha_1, \alpha_2, \ldots, \alpha_k} = \frac{n!}{\alpha_1!\alpha_2!\ldots\alpha_k!}.$$

19. By the multinomial theorem, $(a+b+c)^4 = a^4 + 4a^3b + 6a^2b^2 + 4ab^3 + b^4 + 4a^3c + 12a^2bc + 12ab^2c + 4b^3c + 6a^2c^2 + 12abc^2 + 6b^2c^2 + 4ac^3 + 4bc^3 + c^4$.

20. By the multinomial theorem, the coefficient is $\binom{20}{3,7,10} = 22170720$.

21. (a) $\binom{25}{10,15} = 3268760$

    (b) $\binom{45}{10,15,20} = 10361546974682663760$

◇22. $100!/(10!10!80!)$

    99026143582326261786805320

23. The total number of pizzas is

$$1 + 12 + \binom{2+12-1}{2} + \binom{3+12-1}{3} + \binom{4+12-1}{4} = 1820.$$

24. The number of solutions is $\binom{12}{2} = 66$.

†25. The number of solutions is $\binom{n+k-1}{k-1}$.

26. The problem is equivalent to counting solutions to

$$(2x_1 + 1) + (2x_2 + 1) + \cdots + (2x_n + 1) = k,$$

where the x_i are nonnegative integers. This equation is equivalent to

$$x_1 + x_2 + \cdots + x_n = (k - n)/2,$$

and hence the number of solutions is

$$\binom{n + (k - n)/2 - 1}{(k - n)/2}.$$

Of course, k and n must have the same parity and $k \geq n$.

27. We prove the equivalent identity

$$\sum_{j=0}^{n}\sum_{k=0}^{n} 3\binom{n+j+k}{n,j,k}3^{2n-j-k} = 3^{3n+1}.$$

The right side counts the number of strings of length $3n+1$ composed of symbols A, B, and C. Each such string contains at least $n+1$ As, Bs, or Cs. Reading the string left to right, suppose that the $(n+1)$st occurrence of one symbol takes place in the $(n+j+k+1)$st position, where $0 \le j, k \le n$, and there are already j occurrences of the next available symbol in cyclic order (A, B, C) and k occurrences of the next available symbol in cyclic order. The left side counts the number of strings of this type.

†28. A composition of n can be obtained from a sum of n 1s:

$$n = \underbrace{1 + 1 + \cdots + 1}_{n}.$$

We can place a vertical line between any two consecutive 1s; summing the elements bounded by lines, we obtain a composition. For example,

$$10 = 1 + 1 + 1 + 1 \mid +1 \mid +1 + 1 + 1 \mid +1 + 1$$

gives the composition $10 = 4 + 1 + 3 + 2$. In general, there are $n-1$ places for the vertical lines, so there are 2^{n-1} compositions.

†29. We must show that, for any $n \ge 1$, the number of permutations of $X = \{1, \ldots, n\}$ with an odd number of cycles is equal to the number of permutations of X with an even number of cycles. Clearly, this is true for $n = 1$. Let

$\mathcal{A} = \{$perm. of X with odd # of cycles and 1 in a cycle by itself$\}$

$\mathcal{B} = \{$perm. of X with odd # of cycles and 1 not in a cycle by itself$\}$

$\mathcal{C} = \{$perm. of X with even # of cycles and 1 in a cycle by itself$\}$

$\mathcal{D} = \{$perm. of X with oven # of cycles and 1 not in a cycle by itself$\}$.

Since there are $n-1$ places where 1 can follow one of the other $n-1$ elements, we have

$$(n-1)|\mathcal{A}| = |\mathcal{D}|, \quad (n-1)|\mathcal{C}| = |\mathcal{B}|.$$

The relation $|\mathcal{A}| + |\mathcal{B}| = |\mathcal{C}| + |\mathcal{D}|$ now follows by induction from the hypothesis that $|\mathcal{A}| = |\mathcal{C}|$.

Chapter 4

1. We take successive differences:

$$
\begin{array}{cccccccc}
7, & 11, & 25, & 73, & 203, & 487, & 1021, & \ldots \\
4, & 14, & 48, & 130, & 284, & 534, & \ldots \\
10, & 34, & 82, & 154, & 250, & \ldots \\
24, & 48, & 72, & 96, & \ldots \\
24, & 24, & 24, & \ldots
\end{array}
$$

Having obtained a constant sequence, we stop. We find that the polynomial is

$$
p(n) = 7 + 4n + 10\binom{n}{2} + 24\binom{n}{3} + 24\binom{n}{4} = n^4 - 2n^3 + 4n^2 + n + 7.
$$

2. We see that the $p(n) = n^5$.

3.
$$
n^3 + 2n^2 - n + 1 = 6\binom{n}{3} + 10\binom{n}{2} + 2\binom{n}{1} + 1\binom{n}{0}
$$

4. Define $q(n) = p(2n)$.

†5. Given a polynomial $p(n)$ of degree d, we choose the coefficient of $\binom{n}{d}$ to be $d!$ times the leading coefficient of p. This "kills off" the degree d term of p. We repeat this procedure on the remainder until the entire polynomial is a linear combination of the $\binom{n}{k}$ terms.

†6. The expression n^d counts all functions from $\{1, \ldots, d\}$ to $\{1, \ldots, n\}$. Each such function is onto *some* nonempty subset of $\{1, \ldots, n\}$. The summation counts these onto functions according to the size k of the image.

◇7. One way to find $p(m, n)$ is to use matrices. Let the given matrix be D (for data). Define

$$
P = \begin{bmatrix}
1 & 1 & 1 & 1 & 1 & 1 \\
0 & 1 & 2 & 3 & 4 & 5 \\
0 & 1 & 4 & 9 & 16 & 25 \\
0 & 1 & 8 & 27 & 64 & 125 \\
0 & 1 & 16 & 81 & 256 & 625 \\
0 & 1 & 32 & 243 & 1024 & 3125
\end{bmatrix}.
$$

The columns of P are powers of m, for $0 \le m \le 5$. Then define

$$
C = (P^t)^{-1} D P^{-1}.
$$

Finally,

$$p(m,n) = \sum_{i=1}^{6} \sum_{j=1}^{6} [C]_{ij} m^{i-1} n^{j-1}.$$

With the matrix D of our problem, we obtain

$$p(m,n) = m^3 n^2 + mn + 5m + 6n + 10.$$

Why does this method work?

8. Find the degree of the polynomial that gives this sequence.

9. $3n^3 + 2n^2 + n + 1$

10. Even $p(0) = 1$ and $p(2) = 2$ is impossible.

Chapter 5

1. $(1+x)^{-5} = 1 - 5x + 15x^2 - 35x^3 + 70x^4 - 126x^5 + \cdots$

2. $(1-x^2)^{-4} = 1 + 4x^2 + 10x^4 + 20x^6 + 35x^8 + 56x^{10} + \cdots$

3. Row -6 begins

$$0 \ 1 \ -6 \ 21 \ -56 \ 126 \ -252.$$

4. $49/64$

5. 27

6. $n = 3$

7. If n is positive, then $\binom{n}{4} = 15$, and we see by inspecting Pascal's triangle that the only solution is $n = 6$. If n is negative, then $\binom{-n+4-1}{4} = 15$, and we find that $n = -3$.

◇8.
```
c[n_, -1] := 0; c[0, k_] := 0;
c[0, 0] = 1; c[n_, k_] := c[n, k] =
If[n = 0, c[n-1, k-1] + c[n-1, k], c[n+1, k] - c[n, k-1]];

Table[c[n, k], {n, -10, 10}, {k, 0, 10}] // TableForm
```

Chapter 6

1. The identity holds for $n = 1$, since $F_1 = 1 = 2 - 1 = F_3 - 1$. Assume that the identity holds for n. Then

$$F_1 + \cdots + F_{n+1} = (F_1 + \cdots + F_n) + F_{n+1}$$
$$= F_{n+2} - 1 + F_{n+1}$$
$$= F_{n+3} - 1.$$

Hence, the identity holds for $n + 1$ and by induction for all $n \geq 1$.

2. The identity holds for $n = 1$, since $F_1^2 = 1 = 1 \cdot 1 = F_1 F_2$. Assume that the identity holds for n. Then

$$F_1^2 + \cdots + F_{n+1}^2 = (F_1^2 + \cdots + F_n^2) + F_{n+1}^2$$
$$= F_n F_{n+1} + F_{n+1}^2$$
$$= F_{n+1}(F_n + F_{n+1})$$
$$= F_{n+1} F_{n+2}.$$

Hence, the identity holds for $n + 1$ and by induction for all $n \geq 1$.

3. It follows by induction that

$$\begin{bmatrix} 1 & 1 \\ 1 & 0 \end{bmatrix}^n = \begin{bmatrix} F_{n+1} & F_n \\ F_n & F_{n-1} \end{bmatrix}, \quad n \geq 1.$$

Hence

$$\begin{bmatrix} F_{m+n+1} & F_{m+n} \\ F_{m+n} & F_{m+n-1} \end{bmatrix} = \begin{bmatrix} 1 & 1 \\ 1 & 0 \end{bmatrix}^{m+n}$$
$$= \begin{bmatrix} 1 & 1 \\ 1 & 0 \end{bmatrix}^m \begin{bmatrix} 1 & 1 \\ 1 & 0 \end{bmatrix}^n$$
$$= \begin{bmatrix} F_{m+1} & F_m \\ F_m & F_{m-1} \end{bmatrix} \begin{bmatrix} F_{n+1} & F_n \\ F_n & F_{n-1} \end{bmatrix}.$$

Considering the $(2, 1)$ entry of this matrix identity, we obtain the desired relation.

$\diamond$4. ```
f[0] = 0; f[1] = 1; f[n_] := f[n] = f[n - 1] + f[n - 2];

f[100]
354224848179261915075
```
```

5. The Fibonacci numbers are sums of "shallow diagonals" of Pascal's triangle. For example,

$$1 + 10 + 15 + 7 + 1 = 34.$$

The identity is

$$\binom{n}{0} + \binom{n-1}{1} + \binom{n-2}{2} + \cdots = F_{n+1}, \quad n \geq 0.$$

You can prove the identity by induction.

6. The equation is equivalent to

$$\binom{n+1}{k+1} = \binom{n}{k+2}.$$

Hence, we have

$$\binom{14}{4} + \binom{14}{5} = \binom{14}{6} = 3003.$$

★7. Using Cassini's identity, we have

$$\sum_{n=1}^{\infty} \tan^{-1} \frac{1}{F_{2n+1}} = \sum_{n=1}^{\infty} \tan^{-1} \frac{F_{2n+1}}{F_{2n+1}^2}$$

$$= \sum_{n=1}^{\infty} \tan^{-1} \left(\frac{F_{2n+2} - F_{2n}}{F_{2n} F_{2n+2} + 1} \right)$$

$$= \sum_{n=1}^{\infty} \tan^{-1} \left(\frac{1/F_{2n} - 1/F_{2n+2}}{1 + 1/(F_{2n} F_{2n+2})} \right)$$

$$= \sum_{n=1}^{\infty} \left(\tan^{-1} \frac{1}{F_{2n}} - \tan^{-1} \frac{1}{F_{2n+2}} \right)$$

$$= \tan^{-1} \frac{1}{F_2}$$

$$= \tan^{-1} 1$$

$$= \frac{\pi}{4}.$$

◇8. The number 210 occurs six times in Pascal's triangle:

$$210 = \binom{210}{1} = \binom{210}{209} = \binom{10}{3} = \binom{10}{7} = \binom{16}{2} = \binom{16}{14}.$$

◇9. $\binom{104}{39} = 61218182743304701891431482520$

10. The characteristic polynomial of the sequence $\{a_n\}$ is

$$x^2 - 5x - 6 = (x - 3)(x - 2).$$

Hence

$$a_n = A3^n + B2^n, \quad n \geq 0,$$

for some constants A and B.

From the initial values, $a_0 = 0$ and $a_1 = 1$, we find that

$$A + B = 0$$

$$3A + 2B = 1,$$

and hence $A = 1$, $B = -1$. Therefore

$$a_n = 3^n - 2^n, \quad n \geq 0.$$

11. The characteristic polynomial is

$$x^3 - 3x^2 - 4x + 12,$$

with roots -2, 2, and 3. Hence

$$a_n = A(-2)^n + B2^n + C3^n,$$

for some constants A, B, and C. With the initial values $a_0 = 0$, $a_1 = 1$, and $a_2 = 2$, we find that

$$a_n = -\frac{3}{20}(-2)^n - \frac{1}{4}2^n + \frac{2}{5}3^n, \quad n \geq 0.$$

12. The characteristic polynomial is

$$x^3 - 4x^2 + x + 6,$$

with roots 3, 2, and -1. Hence

$$b_n = A3^n + B2^n + C(-1)^n, \quad n \geq 0,$$

for some constants A, B, and C.

With the initial values $b_0 = 0$, $b_1 = 0$, $b_2 = 1$, we find that

$$b_n = \frac{1}{4}3^n - \frac{1}{3}2^n + \frac{1}{12}(-1)^n, \quad n \geq 0.$$

With the initial values $b_0 = 0$, $b_1 = 1$, $b_2 = 2$, we find that

$$b_n = \frac{1}{4}3^n - \frac{1}{4}(-1)^n, \quad n \geq 0.$$

13. Since $c_n = a_n + b_n$, the characteristic roots of $\{c_n\}$ include the characteristic roots of $\{a_n\}$ and $\{b_n\}$. The characteristic roots of $\{a_n\}$ are 2 and 3. The characteristic roots of $\{b_n\}$ are 4 and 5. Hence, a characteristic polynomial for $\{c_n\}$ is

$$(x-2)(x-3)(x-4)(x-5) = x^4 - 14x^3 + 71x^2 - 154x + 120,$$

and the recurrence for $\{c_n\}$ is

$$c_0 = 1, \; c_1 = 2, \; c_2 = 14, \; c_3 = 80,$$

$$c_n = 14c_{n-1} - 71c_{n-2} + 154c_{n-3} - 120c_{n-4}, \quad n \geq 4.$$

The characteristic roots of $\{d_n\}$ are 8, 10, 12, and 15. Hence, the characteristic polynomial for $\{d_n\}$ is $(x-8)(x-10)(x-12)(x-15) = x^4 - 45x^3 + 746x^2 - 5400x + 14400$. Hence, a recurrence relation is

$$d_n = 45d_{n-1} - 746d_{n-2} + 5400d_{n-3} - 14400d_{n-4}, \quad n \geq 4.$$

14. The characteristic roots of $\{a_n\}$ are ϕ, $\hat{\phi}$, and 2. Hence, the characteristic polynomial is

$$(x^2 - x - 1)(x - 2) = x^3 - 3x^2 + x + 2,$$

and a recurrence relation is

$$a_n = 3a_{n-1} - a_{n-2} - 2a_{n-3}, \quad n \geq 3.$$

15. The answer is the same for (a) and (b). The characteristic polynomial is

$$(x-3)(x-1)^2 = x^3 - 5x^2 + 7x - 3.$$

Hence, the recurrence relation is

$$a_n = 5a_{n-1} - 7a_{n-2} + 3a_{n-3}, \quad n \geq 3.$$

16. We find the particular solution $a_n = -n-3$ to the recurrence relation. Hence, the general solution is of the form

$$a_n = A\phi^n + B\hat{\phi}^n - n - 3.$$

We need to choose A and B so that the initial values are satisfied. Thus,

$$0 = A + B - 3$$

$$1 = A\phi + B\hat{\phi} - 4.$$

We find that

$$A = (3\hat{\phi} - 5)/(\hat{\phi} - \phi) = (15 + 7\sqrt{5})/10$$
$$B = (3\phi - 5)/(\phi - \hat{\phi}) = (15 - 7\sqrt{5})/10.$$

17. We obtain a particular solution assuming it has the form $\alpha 2^n$. Thus

$$\alpha 2^n = \alpha 2^{n-1} + \alpha 2^{n-2} + 2^n,$$

which implies that $\alpha = 4$. Therefore, the solution is

$$a_n = A\phi^n - B\hat{\phi}^n + 4 \cdot 2^n, \quad n \geq 0,$$

for some constants A and B. With the initial values, we find that
$A = (-2\sqrt{5} - 6)/5$ and $B = 2(5 - 3\sqrt{5})/5$.

18. The identity holds for $n = 1$, as $L_1 = 1 = 0 + 1 = F_0 + F_1$. Assume
that the identity holds for $n - 1$ and n. Then

$$\begin{aligned}
L_{n+1} &= L_{n-1} + L_n \\
&= (F_{n-2} + F_n) + (F_{n-1} + F_{n+1}) \\
&= (F_{n-2} + F_{n-1}) + (F_n + F_{n+1}) \\
&= F_n + F_{n+2}.
\end{aligned}$$

Hence, the identity holds for $n + 1$ and by induction for all $n \geq 1$.

19. The identity holds for $n = 1$, as $F_1 = 1 = (2 + 3)/5 = (L_0 + L_1)/5$.
Assume that the identity holds for $n - 1$ and n. Then

$$\begin{aligned}
F_{n+1} &= F_{n-1} + F_n \\
&= (L_{n-2} + L_n)/5 + (L_{n-1} + L_{n+1})/5 \\
&= (L_{n-2} + L_{n-1})/5 + (L_n + L_{n+1})/5 \\
&= (L_n + L_{n+2})/5.
\end{aligned}$$

Hence, the identity holds for $n + 1$ and by induction for all $n \geq 1$.

20. We have

$$F_n L_n = \frac{1}{\sqrt{5}}(\phi^n - \hat{\phi}^n)(\phi^n + \hat{\phi}^n) = \frac{1}{\sqrt{5}}(\phi^{2n} - \hat{\phi}^{2n}) = F_{2n}.$$

21. $L_n^2 - L_{n-1}L_{n+1} = 5(-1)^n, \quad n \geq 1$

22. For any cubic polynomial $p(n)$, we have $0 = \Delta^4 p(n)$, where Δ is the difference operator. Hence

$$0 = p(n+4) - 4p(n+3) + 6p(n+2) - 4p(n+1) + p(n),$$

and a recurrence relation satisfied by all cubic polynomials is

$$a_n = 4a_{n-1} - 6a_{n-2} + 4a_{n-3} - a_{n-4}, \quad n \geq 4.$$

23. We have $a_0 = 0$, $a_1 = 1$, $a_2 = 36$, and $a_3 = 1225$ (the last value perhaps via a computer). The square roots of these numbers are 0, 1, 6, and 35, so we guess that the square roots satisfy the recurrence relation $b_n = 6b_{n-2} - b_{n-2}$, for $n \geq 2$. The characteristic polynomial of this recurrence is $x^2 - 6x + 1$, with roots r_1 and r_2. From the form of the characteristic polynomial, we find that $r_1^2 + r_2^2 = (r_1 + r_2)^2 - 2r_1 r_2 = 6^2 - 2 = 34$ and $r_1^2 r_2^2 = (r_1 r_2)^2 = 1$, so the characteristic polynomial of $\{a_n\}$ is $(x^2 - 34x + 1)(x - 1) = x^3 - 35x^2 + 35x - 1$, and a recurrence relation is $a_n = 35a_{n-1} - 35a_{n-2} + a_{n-2}$, for $n \geq 2$.

24. Let $a_n = 2^n F_n$, for $n \geq 0$. Then $a_0 = 0$, $a_1 = 2$, and $a_n = 2a_{n-1} + 4a_{n-2}$, for $n \geq 2$.

25. Start with the relations

$$F_n = F_{n-1} + F_{n-2}$$

$$F_{n-3} = F_{n-1} - F_{n-2}.$$

Square both equations and add:

$$F_n^2 + F_{n-3}^2 = (F_{n-1} + F_{n-2})^2 + (F_{n-1} - F_{n-2})^2$$

$$= 2F_{n-1}^2 + 2F_{n-2}^2.$$

And so we obtain the recurrence formula

$$F_0^2 = 0, \; F_1^2 = 1, \; F_2^2 = 1, \quad F_n^2 = 2F_{n-1}^2 + 2F_{n-2}^2 - F_{n-3}^2, \quad n \geq 3.$$

$\star$26. The characteristic polynomial of the recurrence relation for the kth powers of the Fibonacci numbers is

$$\prod_{i=0}^{\lfloor (k-1)/2 \rfloor} (x^2 + (-1)^{i+1} L_{k-2i} x + (-1)^k) \cdot \begin{cases} 1 & \text{if } k \bmod 4 = 1, 3 \\ (x-1) & \text{if } k \bmod 4 = 0 \\ (x+1) & \text{if } k \bmod 4 = 2, \end{cases}$$

where L_n is the nth Lucas number.

Chapter 7

1. $(xf'(x))|_{x=1/3} = 6/5$

2. The generating function is $(2 - x)/(1 - x - x^2)$.

3. The denominator is $1 - 5x + 6x^2$. To find the numerator, we multiply $x(1 - 5x + 6x^2)$, keeping only those terms of degree less than 2; so the numerator is x. Hence, the generating function is $x/(1 - 5x + 6x^2)$.

4. We put $x = -1/10$ in the generating function and obtain the value $-5/78$.

5. Define $\{a_n\}$ by $a_0 = 1$, $a_1 = -3$, and $a_n = -3a_{n-1} - a_{n-2}$, for $n \geq 2$.

6. While we can give a proof by induction, a fast proof uses generating functions. Since

$$\sum_{n=0}^{\infty} F_n x^n = \frac{x}{1 - x - x^2},$$

we have

$$\sum_{n=0}^{\infty} (-1)^n F_{2n+2} x^{2n+2} = \frac{1}{2} \left(\frac{x}{1 - x - x^2} + \frac{-x}{1 + x - x^2} \right) = \frac{x^2}{1 - 3x^2 + x^4},$$

and the desired formula follows.

7. Define $\{a_n\}$ by $a_0 = 1$, $a_1 = -1$, and $a_n = -a_{n-1} - 2a_{n-2}$, for $n \geq 2$.

8. $x(x((1 - x)^{-1})')' = x(x + 1)(1 - x)^{-3}$

◇9. Series[
    1/((1-x)(1-x^5)(1-x^10)(1-x^25)(1-x^50)(1-x^100)),
    {x, 0, 100}]

    ...+50x^50+...

◇10. The generating function

$$((1 + x)(1 + x^5)(1 + x^{10})(1 + x^{25})(1 + x^{50})(1 + x^{100}))^{-1}$$

counts the difference between the number of ways of making change with an even number of coins and an odd number of coins. The coefficient of x^{100} is 19. Let e and o be the number of ways of making change for \$1 with an even number of coins and an odd number of coins, respectively. Then $e - o = 19$ and $e + o = 293$. Hence $e = 156$.

11. The coefficients of the generating function for the difference between sums with an odd number of coins and sums with an even number of coins alternate between positive and negative. Surprisingly, there is a one-line proof of this fact:

$$\frac{1}{(1+x)(1+x^5)(1+x^{10})(1+x^{25})(1+x^{50})(1+x^{100})}$$
$$= \frac{(1-x)(1-x^5)(1-x^{25})}{(1-x^2)(1-x^{20})(1-x^{200})}.$$

We need to interpret this equation to see that it proves that the coefficients alternate in sign. The numerator is a polynomial with alternating coefficients, while the denominator is equal to an infinite series with only positive coefficients of even powers of x. Hence the resulting series has alternating coefficients.

Is the condition $2k \in S \implies k \in S$ necessary as well as sufficient?

12. We fill in the following table by hand.

	P_n	N_n	D_n	Q_n	H_n	W_n
5	1	1	0	0	0	0
10	1	2	1	0	0	0
15	1	3	2	0	0	0
20	1	4	4	0	0	0
25	1	5	6	1	0	0
30	1	6	9	2	0	0
35	1	7	12	4	0	0
40	1	8	16	6	0	0
45	1	9	20	9	0	0
50	1	10	25	13	1	0
54	1	11	30	18	2	0
60	1	12	36	24	4	0
65	1	13	42	31	6	0
70	1	14	49	39	9	0
75	1	15	56	49	13	0
80	1	16	64	60	18	0
85	1	17	71	73	25	0
90	1	18	81	87	31	0
95	1	19	89	103	39	0
100	1	20	100	121	50	1

We conclude that the number of ways to make change for a dollar is

$$P_{100} + N_{100} + D_{100} + Q_{100} + H_{100} + W_{100} = 293.$$

13. (a)

$$\frac{1}{1-x-y} = \frac{1}{1-(x+y)} = 1 + (x+y) + (x+y)^2 + (x+y)^3 + \cdots$$

(b)

$$\frac{1}{1-x-y-z} = \frac{1}{1-(x+y+z)} = 1 + (x+y+z) + (x+y+z)^2 + (x+y+z)^3 + \cdots$$

◇14. We find that

$$\frac{1}{(1-x)(1-x^2)(1-x^4)}$$

$$= \frac{1}{8(1-x)^3} + \frac{1}{4(1-x)^2} + \frac{9}{32(1-x)} + \frac{1}{16(1+x)^2} + \frac{5}{32(1+x)} + \frac{1+x}{8(1+x^2)}.$$

The coefficient of x^n is

$$\frac{1}{8}\binom{-3}{n}(-1)^n + \frac{1}{4}\binom{-2}{n}(-1)^n + \frac{9}{32} + \frac{1}{16}\binom{-2}{n} + \frac{5}{32}(-1)^n$$

$$+ \left\{ \begin{array}{ll} \frac{1}{8}(-1)^{n/2} & n \text{ even} \\ \frac{1}{8}(-1)^{(n-1)/2} & n \text{ odd} \end{array} \right\}.$$

Using the identity $\binom{-\alpha}{n} = (-1)^n \binom{\alpha+n-1}{n}$, this simplifies to

$$\frac{(n+2)(n+1)}{16} + \frac{n+1}{4} + \frac{9}{32} + \frac{(n+1)(-1)^n}{16} + \frac{5}{32}(-1)^n$$

$$+ \left\{ \begin{array}{ll} \frac{1}{8}(-1)^{n/2} & n \text{ even} \\ \frac{1}{8}(-1)^{(n-1)/2} & n \text{ odd} \end{array} \right\}.$$

Hence, the number of solutions with $n = 10^{30}$ is

625000000000000000000000000000005000000000000000000000000000001.

◇15. We find that

$$\frac{1}{(1-x)(1-x^2)(1-x^3)}$$

$$= \frac{1}{6(1-x)^3} + \frac{1}{4(1-x)^2} + \frac{17}{72(1-x)} + \frac{1}{8(1+x)} + \frac{2+x}{9(1+x+x^2)}.$$

The coefficient of x^n is

$$\frac{1}{6}\binom{-3}{n}(-1)^n + \frac{1}{4}\binom{-2}{n}(-1)^n + \frac{17}{72} + \frac{1}{8}(-1)^n + (\text{pattern mod } 3).$$

This simplifies to

$$\frac{(n+2)(n+1)}{16} + \frac{n+1}{4} + \frac{17}{72} + \frac{1}{8}(-1)^n + (\text{pattern mod } 3).$$

Plugging in $n = 10^{30}$, we find that the number of solutions is

$$\frac{1}{6}(10^{32} + 2)(10^{30} + 1) + \frac{1}{4}(10^{30} + 1) + \frac{17}{72} + \frac{1}{8} - \frac{1}{9} = 8\underbrace{3\ldots3}_{57}4.$$

This is equal to the number of partitions of 10^{30} into 1, 2, or 3 parts:

$$1 + \frac{10^{30}}{2} + \left\{\frac{(10^{30})^2}{12}\right\}.$$

◇16. We find that

$$\frac{1}{(1-x)(1-x^2)(1-x^4)} = \frac{1}{12(1-x)^3} + \frac{5}{24(1-x)^2}$$

$$+ \frac{41}{144(1-x)} + \frac{1}{16(1+x)} + \frac{1}{4(1+x^2)} + \frac{1+2x}{9(1+x+x^2)}.$$

The coefficient of x^n is

$$\frac{1}{12}\binom{-3}{n}(-1)^n + \frac{5}{24}\binom{-2}{n}(-1)^n + \frac{41}{144} + \frac{1}{16}(-1)^n$$

$$+ (\text{pattern mod } 4) + (\text{pattern mod } 3).$$

This simplifies to

$$\frac{(n+2)(n+1)}{24} + \frac{5(n+1)}{24} + \frac{41}{144} + \frac{1}{16}(-1)^n$$

$$+ (\text{pattern mod } 4) + (\text{pattern mod } 3).$$

With $n = 10^{30}$, we have

$$4\underbrace{6\ldots6}_{26}7\underbrace{0\ldots0}_{29}1$$

solutions.

⋆17. Let E be the operator xD_x. We will show that $E^k(1+x)^n$ is a function of the form $(1+x)^{n-k}q(x,n)$, where q is a polynomial in x and n, of degree k in x, with leading term $n^k x^k$. The result then follows upon letting $x = 1$. The binomial theorem says that

$$\sum_{i=1}^{n}\binom{n}{i}x^i = (1+x)^n,$$

so the claim is true for $k = 0$. Assume that it is true for k. Then

$$E^{k+1}(1+x)^n = xD_x(1+x)^{n-k}q(x,n)$$

$$= x(n-k)(1+x)^{n-k-1}q(x,n) + x(1+x)^{n-k}q'(x,n)$$

$$= (1+x)^{n-k-1}[x(n-k)q(x,n) + x(1+x)q'(x,n)].$$

The leading term of the second factor is

$$x^{k+1}n^k(n-k) + x^2 kx^{k-1}n^k = n^{k+1}x^{k+1}.$$

Hence, the result is true for $k+1$ and by induction for all $k \geq 0$. Upon plugging in $x = 1$, we obtain the desired result.

18. Let

$$u = \sum_{n=0}^{\infty} n!x^n + \sum_{n=0}^{\infty} (2x)^n$$

$$= \sum_{n=0}^{\infty} n!x^n + \frac{1}{1-2x}.$$

Then

$$ux = \sum_{n=0}^{\infty} n!x^{n+1} + \frac{x}{1-2x}$$

and so

$$(ux)' = u'x + u = \sum_{n=0}^{\infty} (n+1)!x^n + \frac{1}{(1-2x)^2}.$$

Hence

$$u'x^2 + ux = \sum_{n=0}^{\infty} (n+1)!x^{n+1} + \frac{x}{(1-2x)^2}.$$

It follows that

$$u'x^2(2x-1)^2 + u(x-1)(2x-1)^2 = g(x),$$

where $g(x)$ is a polynomial. Upon identifying the coefficient of x^n, we obtain the recurrence relation

$$a_n = (n+4)a_{n-1} - 4na_{n-2} + (4n-8)a_{n-3}, \quad n \geq 3.$$

★19. The generating function that indicates all positive integers is $1/(1-x)$. Find the generating function that indicates amounts that can be made and subtract from $1/(1-x)$. Show that the result is a polynomial and find its degree and number of nonzero terms.

◇20.
$$13333398333445333413833354500001 \doteq 1.3 \times 10^{31}$$

◇⋆21.
$$40125046347199679029952380920610239599328534571 30267501 \doteq 4.0 \times 10^{54}$$

Chapter 8

1. 20833

◇2. The partial fractions decomposition is
$$-\frac{1}{24(x-1)^3} + \frac{13}{288(x-1)} - \frac{1}{16(x+1)^2} - \frac{1}{32(x+1)}$$
$$-\frac{x+1}{8(x^2+1)} + \frac{x+2}{9(x^2+x+1)}.$$
Use binomial series.

3. The triangles $(4,4,7)$ and $(1,2,2)$ both have an angle with cosine $7/8$. (The triangle $(2,3,4)$ has the same angle.) The triangles $(3,7,8)$ and $(5,7,8)$ both have an angle of $60°$ and hence cosine $1/2$.

◇4. Use the law of cosines.

5. Use the law of cosines.

◇6. $(11, 39, 49)$

7. Let $s(n)$ be the number of scalene integer triangles of perimeter n. Then $s(n) = t(n-6)$, for $n \geq 6$. Let (a,b,c) be a scalene triangle and consider the triangle $(a-1, b-2, c-3)$. Alternatively, find the generating function by building up from the smallest scalene triangle.

8. We will show that for $k \geq 3$, there are more triangles with (odd) perimeter $2n+1$ than triangles with (even) perimeter $2k$ or $2k+2$. Suppose that a triangle with perimeter $2k$ has sides a, b, c, with $a \leq b \leq c$. Then the triangle with sides a, b, $c+1$ has perimeter $2k+1$, and it is a genuine triangle because $a+b = c+1$ is impossible since $2k+1$ is odd. Also, the triangle with sides $1, k, k$ does not come from such a transformation, so $t(2k+1)$ is strictly greater than $t(2k)$. Similarly, starting with a triangle with sides a, b, c, where $a \leq b \leq c$, and $a+b+c = 2k+2$, we find that the triangle with sides a, b, and $c-1$ satisfies the triangle inequality. Also, the triangle with sides $1, k, k$ does not come from such a transformation, so $t(2k+1) > t(2k+2)$. Notice that the restriction $k \geq 3$ is needed to ensure that the triangle with sides $1, k, k$ isn't the only triangle with perimeter $2k+1$.

9. The only such number is $n = 48$.

10. We can find a simple formula for $t(n)$, where n has any given remainder upon division by 12. Thus

$$t(12k) = 3k^2$$

$$t(12k + 1) = 3k^2 + 2k$$

$$t(12k + 2) = 3k^2 + k$$

$$t(12k + 3) = 3k^2 + 3k + 1$$

$$t(12k + 4) = 3k^2 + 2k$$

$$t(12k + 5) = 3k^2 + 4k + 1$$

$$t(12k + 6) = 3k^2 + 3k + 1$$

$$t(12k + 7) = 3k^2 + 5k + 2$$

$$t(12k + 8) = 3k^2 + 4k + 1$$

$$t(12k + 9) = 3k^2 + 6k + 3$$

$$t(12k + 10) = 3k^2 + 5k + 2$$

$$t(12k + 11) = 3k^2 + 7k + 4.$$

Since these formulas are all quadratic polynomials, $t(n)$ satisfies the recurrence relation

$$t(12k + r) = 3t(12(k - 1) + r) - 3t(12(k - 2) + r) + t(12(k - 3) + r).$$

(See Exercise 22 of Chapter 6.) The desired recurrence relation follows immediately.

⋆11. First, we prove that the period of $\{t(n) \bmod m\}$ is at most $12m$. If n is even, we may write $n = 12m + 2r$ and we have

$$t(12m + 2r) = \{(12m + 2r)^2/48\}$$

$$= 3m^2 + mr + \{(2r)^2/48\}$$

$$\equiv t(2r) \pmod{m}.$$

The odd case is similar, so that $t(12m + u) \equiv t(u) \bmod m$, for all u. Hence, the period of $\{t(n) \bmod m\}$ is a divisor of $12m$ and therefore at most $12m$.

Second, we show that the period of $\{t(n) \bmod m\}$ is at least $12m$. Suppose that the period is p. Then let $k = p$ or $p - 1$ so that k is even. We then have $\{k^2/48\} \equiv 0 \pmod{m}$ and $\{(k + 2)^2/48\} \equiv 0 \pmod{m}$ since $t(-1) = t(0) = t(1) = t(2) = 0$. Thus, m divides $\{(k + 2)^2/48\} - \{k^2/48\}$, which is nonzero because $p > 12$. This difference is less than $(k + 2)^2/48 - k^2/48 + 1 = (k + 13)/12$. Hence $12m < k + 13 \leq p + 13$, and this completes the proof since $12m$ is a multiple of p and $p > 12$.

Note. Every linear recurrence sequence is periodic with respect to any given modulus. The period is the least common multiple of the periods with respect to the prime power divisors of the modulus. It is surprising that for the well-studied Fibonacci sequence, no formula is known for the periods with respect to prime power moduli; however, it is known that the period modulo 2^n is $3 \cdot 2^{n-1}$ and the period modulo 5^n is $4 \cdot 5^n$.

12. One such infinite family of pairs is $(2n, 4n^2 - 3n + 1, 4n^2 - 3n + 1)$ and $(2n^2 - n + 1, 2n^2 - n + 1, 4n^2 - 2n)$, where $n \geq 2$, with common perimeter $8n^2 - 4n + 2$ and common area $n(2n - 1)\sqrt{4n^2 - 2n + 1}$.

Chapter 9

1. The recurrence formula is

$$k(1, 1) = 1;$$

$$k(0, n) = 0; \quad k(m, 0) = 0;$$

$$k(m, n) = k(m - 1, n) + k(m, n - 1) + k(m - 1, n - 1), \quad m, n \geq 1.$$

We calculate $k(8, 8) = 48639$.

2. This is the number of compositions (ordered partitions) of n. See p. 18.

◇3. 75059524392.

◇4.

0	0	0	-2	1
0	0	3	1	-2
0	0	-4	3	0
0	0	0	0	0
0	0	0	0	0

0	0	3	1	-2
0	-4	-2	3	1
0	1	-4	-2	3
0	5	1	-4	0
0	0	0	0	0

0	0	-4	3	0
0	1	-4	-2	3
5	5	4	-4	-4
-6	-2	5	1	0
0	-6	5	0	0

0	0	0	0	0
0	5	1	-4	0
-6	-2	5	1	0
1	-6	-2	5	0
7	1	-6	0	0

0	0	0	0	0
0	0	0	0	0
0	-6	5	0	0
7	1	-6	0	0
-8	7	0	0	0

The number of Queen paths from $(0,0,0)$ to $(7,7,7)$ is $1,540,840,801,552$.

5. The generating function is

$$\frac{(1-x)(1-xy)}{1-2(x+xy)+3(x \cdot xy)}.$$

Hence, the recurrence formula is

$$a(m,n) = 2a(m-1,n) + 2a(m-1,n-1) - 3a(m-2,n-1).$$

The initial values are $a(0,0) = 1$, $a(1,0) = 1$, $a(2,0) = 2$, $a(1,1) = 1$, and $a(2,1) = 4$.

◇6. We have

$$a(n) = 2a(n-1) + 2a(n-2) - 3a(n-3), \quad n \geq 4,$$

$$a(0) = 1, \, a(1) = 1, \, a(2) = 3, \, a(4) = 6.$$

With the help of a computer, we find that

$$a(100) = 870338141873214655919573200648700175$$

$$\doteq 8.7 \times 10^{35}.$$

7. The only value is $n = 3$. To prove uniqueness, consider the sequence modulo 3.

★8.

$$\sum\sum a(m,n)x^m y^n = \frac{1}{1 - (x/(1-x)) - (y/(1-y))}$$

$$= \sum_{k=0}^{\infty} \left[\left(\frac{x}{1-x}\right) + \left(\frac{y}{1-y}\right) \right]^k$$

$$= \sum_{p=0}^{\infty}\sum_{q=0}^{\infty} \binom{p+q}{p} \left(\frac{x}{1-x}\right)^p \left(\frac{y}{1-y}\right)^q$$

$$a(m,n) = \sum_{p=0}^{m}\sum_{q=0}^{n} \binom{p+q}{p}\binom{m-1}{p-1}\binom{n-1}{q-1}$$

★9. (a) The recurrence formula is

$$a(0,0) = 1; \, a(0,1) = 1; \, a(1,0) = 1;$$

$$a(-1,n) = 0; \, a(-2,n) = 0; \, a(m,-1) = 0; \, a(m,-2) = 0;$$

$$a(m,n) = a(m-1,n) + a(m-2,n) + a(m,n-1) + a(m,n-2),$$

$$m,n \geq 2.$$

(b) The generating function is

$$\frac{1}{1 - s - t - s^2 - t^2}.$$

(c) Set $st = x$ and find s^0.

⋆10. (a) The recurrence formula is

$b(0,0) = 1; \; b(0,1) = 1; \; b(1,0) = 1; \; b(1,1) = 3;$

$b(2,0) = 2; \; b(0,2) = 2; \; b(2,1) = 7; \; b(1,2) = 7; \; b(2,2) = 21;$

$b(m,-1) = 0; \; b(m,-2) = 0; \; b(-1,n) = 0; \; b(-1,n) = 0;$

$b(m,n) = 2b(m-1,n) + 2b(m,n-1) - 2b(m-1,n-1)$

$\qquad - b(m-2,n-1) - b(m-1,n-2) + b(m-2,n-2), \quad m,n \geq 2.$

(b) The denominator of the generating function is

$$1 - 2s - 2t + 2st + s^2t + st^2 - s^2t^2.$$

(c) Set $st = x$ and find s^0.

⋆11. (a) The recurrence formula is

$c(0,0) = 1; \; c(0,1) = 1; \; c(1,0) = 1;$

$c(-1,n) = 0; \; c(-2,n) = 0; \; c(m,-1) = 0; \; c(m,-2) = 0;$

$c(m,n) = 2c(m-1,n) - c(m-1,n-1) + c(m,n-1),$

$\qquad m,n \geq 2.$

(b) The denominator of the generating function is

$$1 - 2s - t + st.$$

(c) Set $st = x$ and find s^0.

⋆12. (a) The recurrence formula is

$d(0,0) = 1; \; d(0,1) = 1; \; d(1,0) = 1; \; d(1,1) = 3;$

$d(-1,n) = 0; \; d(m,-1) = 0;$

$d(m,n) = d(m-1,n) + d(m,n-1)$

$\qquad + 2d(m-1,n-1) - d(m-2,n-1) - d(m-1,n-2), \quad m,n \geq 2.$

(b) The generating function is

$$\frac{1 - st}{1 - s - t - 2st + s^2t + st^2}.$$

(c) Set $st = x$ and find s^0.

13. The generating function is

$$\frac{1}{1 - t(x/(1-x)) - t(y/(1-y))} = \frac{(1-x)(1-y)}{1 - x - tx - y - ty + xy + 2txy}.$$

We read off the recurrence relation:

$$a(m, n; k) = a(m-1, n; k) + a(m-1, n; k-1) + a(m, n-1; k)$$
$$+ a(m, n-1; k-1) - a(m-1, n-1; k) - 2a(m-1, n-1; k-1).$$

14. A generalized Knight path from $(0,0)$ to (m,n) is possible if and only if there exist nonnegative integers α and β such that $\alpha(1,2)+\beta(2,1) = (m,n)$. Solve this system for α and β. Then find a connection between generalized Knight paths and Rook paths.

◇15. The number of Nim games is equal to the number of Rook paths from $(0,0,0)$ to $(10,10,20)$.

◇16. The recurrence formula is

$$a_0 = 1,\ a_1 = 6,\ a_2 = 222,\ a_3 = 9918;$$

$$0 = (2n^3 - 2n^2)a_n + (-121n^3 + 212n^2 - 85n - 6)a_{n-1}$$
$$+ (-475n^3 + 3462n^2 - 7853n + 5658)a_{n-2}$$
$$+ (1746n^3 - 14580n^2 + 40662n - 37908)a_{n-3}$$
$$+ (-1152n^3 + 12672n^2 - 46080n + 55296)a_{n-4},\quad n \geq 4.$$

This result is empirical; we don't have a proof of it.

Chapter 10

1. The sample space consists of all partitions of 52 cards into two subsets of size 26. The number of such partitions is $\binom{52}{26}$. The probability of each simple event is $1/\binom{52}{26}$.

2. The sample space consists of all ordered pairs of numbers between 1 and 6. The probability that the sum of the two dice is 7 is $6/36 = 1/6$.

3. The probability that the sum is at least 17 is $(1+3)/216 = 1/54$.

4.

$$\Pr\left(X < \frac{n}{2}\right) = \left(\frac{1}{2}\right)^n \sum_{k=0}^{\lfloor n/2 \rfloor} \binom{n}{k} = \left(\frac{1}{2}\right)^n 2^{n-1} = \frac{1}{2}$$

5. The expected number of flips is

$$1q + 2pq + 3p^2q + 4p^3q + 5p^4q + \cdots = q(1 + 2p + 3p^2 + 4p^3 + 5p^4 + \cdots).$$

From the series $(1-x)^{-1} = 1+x+x^2+x^3+\cdots$, we obtain $((1-x)^{-1})' = (1-x)^{-2} = 1 + 2x + 3x^2 + 4x^3 + \cdots$. Hence, the expected number of flips is $p(1-q)^{-2} = 1/p$.

6. For $2 \leq i \leq n$, let X_i be a random variable equal to 1 if the ith element in the string begins a run, and 0 otherwise. Then, for each i, we have $E(X_i) = pq+qp = 2pq$ (conditioning on the previous element). Hence, the expected value of the sum of the X_i is $2pq(n-1)$. Since the first element in the string begins a run, the expected number of runs is $2pq(n-1) + 1$.

7. For $1 \leq i \leq 365$, let $X_i = 1$ if date i is among the birth dates. Then each $E(X_i) = 1 - (364/365)^n$, so that $E(X_1 + \cdots + X_{365}) = 365(1 - (364/365)^n)$.

†8. A recurrence formula for $\{d_n\}$ is

$$d_0 = 1, \ d_1 = 0; \ d_n = (n-1)(d_{n-1} + d_{n-2}), \quad n \geq 2.$$

In a derangement of $\{1, 2, 3, \ldots, n\}$, the element n occurs in a cycle of length 2 or in a cycle of greater length. In a cycle of length 2, there are $n-1$ choices for the other element, while the remaining elements constitute a derangement of $n-2$ elements. In a cycle of length greater than 2, there are $n-1$ choices for the element that precedes n, while the elements other than n constitute a derangement of $n-1$ elements.

◇9. d[0] = 1; d[1] = 0;
d[n_] := d[n] = (n - 1)(d[n - 1] + d[n - 2]);
d[30]

9758107383683577773237742835481

†10. The formula holds for $n = 2$, since $d_2 = 1 = 2d_1 - 1$. Assume that the formula holds for n. Then

$$d_{n+1} = n(d_{n-1} + d_n)$$
$$= nd_{n-1} + nd_n$$
$$= d_n - (-1)^n + nd_n$$
$$= (n+1)d_n + (-1)^{n+1}.$$

Hence, the formula holds for $n + 1$ and by induction for all $n \geq 2$.

The recurrence relation implies the formula $d_n = \sum_{j=0}^{n}(-1)^j n!/j!$.

†⋆11. Let

$$d(x) = \sum_{n=0}^{\infty} d_n \frac{x^n}{n!}.$$

From a recurrence relation for $\{d_n\}$, we find that

$$(1 - x)d'(x) = xd(x).$$

We solve this differential equation using separation of variables:

$$\int \frac{d'(x)}{d(x)}\, dx = \int \frac{x}{1 - x}\, dx$$

and hence

$$d(x) = C\frac{e^{-x}}{1 - x},$$

for some constant C. The condition $d_0 = 1$ implies that $C = 1$. Therefore

$$d(x) = \frac{e^{-x}}{1 - x}.$$

⋆12.

$$
\begin{aligned}
E &= \frac{1}{n!}\sum_{k=1}^{n} k\binom{n}{k} d_{n-k} \\[2mm]
&= \sum_{k=1}^{n} \frac{k}{k!(n-k)!} \sum_{j=0}^{n-k}(-1)^j \frac{(n-k)!}{j!} \\[2mm]
&= \sum_{k=0}^{n-1}\sum_{j=0}^{n-k-1} \frac{(-1)^j}{k!j!} \\[2mm]
&= \sum_{t=0}^{n-1}\sum_{k+j=t} \frac{(-1)^j}{k!j!} \\[2mm]
&= 1 + \sum_{t=1}^{n-1}\frac{1}{t!}\sum_{j=0}^{t}(-1)^j \binom{t}{j} \\[2mm]
&= 1 + \sum_{t=1}^{n-1}\frac{1}{t!}\cdot 0 \\[2mm]
&= 1
\end{aligned}
$$

13. Let X_i, for $1 \le i \le 52$, be 1 if the ith card in the two decks matches and 0 otherwise. Then the expected number of matches is

$$E(X_1 + \cdots + X_n) = E(X_1) + \cdots + E(X_n) = \frac{1}{n} + \cdots + \frac{1}{n} = 1.$$

$\star$14. The probability of exactly two players having a hand with all cards the same suit is

$$\binom{4}{2}\binom{4}{2} 2 \left(\binom{26}{13} - 2 \right) \Big/ \left(\binom{52}{13}\binom{39}{13}\binom{26}{13}\binom{13}{13} \right),$$

(as there are two non-allowed hands for the two players that don't have a hand with all cards the same suit). It is not possible for exactly three players to have all cards of the same suit. Hence, the probability of exactly one player having a hand with all cards the same suit is

$$\frac{18772910672458601}{74506580229845545461005200000} - \frac{1733433}{12417763371640924268342000 0}$$

$$- \frac{24}{\binom{52}{13}\binom{39}{13}\binom{26}{13}} = \frac{242753155112819}{96344715814455446909550000}.$$

We can also solve the problem by defining six events $E_{i,j}$, where $1 \le i < j \le 4$, equal to the probability that players i and j both have perfect hands, and using the inclusion–exclusion principle.

$\dagger\star$15. Suppose that the factorization of n into prime powers is $n = \prod_{i=1}^{k} p_i^{\alpha_i}$. For $1 \le i \le k$, let $X_i = \{y : 1 \le y \le n \text{ and } p_i \mid y\}$. Then

$$
\begin{aligned}
\phi(n) &= n - |X_1 \cup \cdots \cup X_k| \\
&= n - \left(\frac{n}{p_1} + \frac{n}{p_2} + \cdots \right) + \left(\frac{n}{p_1 p_2} + \frac{n}{p_1 p_3} + \cdots \right) - \cdots \\
&= n \left(1 - \frac{1}{p_1} \right) \cdots \left(1 - \frac{1}{p_k} \right).
\end{aligned}
$$

$\star$16. Using Stirling's estimate,

$$n! \sim n^n e^n \sqrt{2\pi n},$$

we obtain

$$\binom{2n}{n} = \frac{(2n)!}{n! n!} \sim \frac{2^{2n}}{\sqrt{2\pi n}},$$

from which the result follows directly.

⋆17. Assume that die P takes the values 1, 2, 3, 4, 5, 6 with probabilities p_1, p_2, p_3, p_4, p_5, p_6, respectively, and die Q takes the values 1, 2, 3, 4, 5, 6 with probabilities q_1, q_2, q_3, q_4, q_4, q_5, q_6, respectively. Then $p_1q_1 = 1/11$ and $p_6q_6 = 1/11$. By the arithmetic mean–geometric mean (AM–GM) inequality,

$$\frac{p_1q_6 + p_6q_1}{2} \geq \sqrt{p_1q_6p_6q_1} = \sqrt{p_1q_1p_6q_6} = \frac{1}{11},$$

and hence

$$p_1q_6 + q_1p_6 \geq \frac{2}{11},$$

contradicting the fact that $p_1q_6 + p_2q_5 + p_3q_4 + p_4q_3 + p_5q_2 + p_6q_1 = 1/11$. Therefore, not all sums occur with the same probability.

⋆18. It is impossible. Let the probabilities of the dies coming up i, for $1 \leq i \leq 6$, be p_i, q_i, r_i, respectively. Then if all the sums are equal, $p_1q_1r_1 = 1/16$ and $p_6q_6r_6 = 1/16$. By the arithmetic mean–geometric mean inequality,

$$\frac{p_1q_6r_1 + p_1q_1r_6 + p_6q_1r_1}{3} \geq (p_1^2q_1^2r_1^2p_6q_6r_6)^{1/3} = \frac{1}{16}.$$

and this contradicts the fact that $p_1q_6r_1 + p_1q_1r_6 + p_6q_1r_1 \leq 1/16$.

19. $\binom{9}{8}(0.9)^8(0.1)(0.9) = 0.348678$

20.
$$\frac{\binom{10}{5}\binom{20}{5}}{\binom{30}{10}} = \frac{62016}{476905} \doteq 0.130038$$

◇21.

$$\sum_{k=48}^{52} \frac{\binom{100}{k}\binom{100}{100-k}}{\binom{200}{100}}$$

$$= \frac{9357465682292947654228528899123969729907841641 25892}{17981242060938581258397839903571334589182 6556075415}$$

$$\doteq 0.520402$$

22.
$$\frac{\binom{9}{3}\binom{9}{3}}{\binom{27}{9}} = \frac{197568}{1562275} \doteq 0.126462$$

23. The expected number of balls remaining in the urn is

$$\frac{w}{b+1} + \frac{b}{w+1}.$$

The proof is a straightforward generalization of the argument in Example 10.19.

24. Let P_n be the probability that, when three balls are chosen, the sum of their numbers is divisible by 3. Note that $P_1 = 1$ and $P_2 = 1/4$. We must show that $P_n \geq 1/4$, for all $n \geq 3$. The only relevant property of the integers $1, \ldots, n$ is their residues modulo 3. Thus, when $n \geq 3$, we may think of there being three integers, 0, 1, 2, occurring in the urn with certain probabilities. Let i be an integer selected at random. We consider three cases:

(1) $n = 3k$. Here $\Pr(i = 0) = k/n$, $\Pr(i = 1) = k/n$, and $\Pr(i = 2) = k/n$. The only triples of residues that sum to 0 modulo 3 are 000, 111, 222, and 012. Hence

$$P_n = \frac{3k^3 + 3k^3 + 3k^3 + 6k^3}{n^3} = \frac{9k^3}{n^3} = \frac{1}{3}.$$

(2) $n = 3k + 1$. Here $\Pr(i = 0) = k/n$, $\Pr(i = 1) = (k+1)/n$, and $\Pr(i = 2) = k/n$. Hence

$$P_n = \frac{k^3 + (k+1)^3 + k^3 + 6k^2(k+1)}{n^3} = \frac{9k^3 + 9k^2 + 3k + 1}{27k^3 + 27k^2 + 9k + 1} > \frac{1}{3}.$$

(3) $n = 3k + 2$. Here $\Pr(i = 0) = k/n$, $\Pr(i = 1) = (k+1)/n$, and $\Pr(i = 2) = (k+1)/n$. Hence

$$P_n = \frac{k^3 + (k+1)^3 + (k+1)^3 + 6k(k+1)^2}{n^3} = \frac{9k^3 + 18k^2 + 12k + 2}{27k^3 + 54k^2 + 18k + 8}$$

$$= \frac{1}{4} + \frac{9k^3 + 18k^2 + 30k}{108k^3 + 216k^2 + 72k + 32} > \frac{1}{4}.$$

25. We can think of this process as waiting time until a success occurs. At first, the probability of a success (withdrawing a black ball) is $5/(5 + 5)$. The average waiting time is therefore $10/5$. After one black ball has been withdrawn, then the probability of success is $4/9$ and the average waiting time $9/4$, etc. The total average waiting time is therefore

$$\frac{10}{5} + \frac{9}{4} + \frac{8}{3} + \frac{7}{2} + \frac{6}{1} = \frac{197}{12}.$$

26. The desired outcomes culminate with a ball chosen from Urn A. The number of balls selected from Urn B is some integer k between 0 and 4, meaning that $5 + k$ balls are selected altogether. With $5 + k$ selections of the urns, the probability of selecting Urn A 5 times and Urn B k times, with Urn A selected last, is $(1/2)^{4+k}\binom{4+k}{k}(1/2)$. If k balls are selected from Urn B, then the probability that the black ball is not selected is $(5 - k)/5$. Therefore, the probability of a desired outcome is

$$\sum_{k=0}^{4}\binom{4+k}{k}\left(\frac{1}{2}\right)^{5+k}\left(\frac{5-k}{5}\right) = \frac{63}{256}.$$

Chapter 11

1. The steady-state solution is $[x, y, z] = [16/53, 15/53, 22/53]$.

eigenvector	eigenvalue
$[16/53, 15/53, 22/53]$	1
$[2 - \sqrt{6}, -3 + \sqrt{6}, 1]$	$-(6 + \sqrt{6})/12$
$[2 + \sqrt{6}, -3 - \sqrt{6}, 1]$	$(-6 + \sqrt{6})/12$

The matrix P^∞ consists of three columns equal to the eigenvector corresponding to eigenvalue 1.

◇2. `m={{0,1/3, 1/2},{1/4, 0, 1/2},{3/4, 2/3, 0}};`
`MatrixPower[m,100].{{1},{0},{0}}-{{16/53},{15/53},{22/53}}`
`//N`

3. The given matrix is a permutation matrix; therefore, it is not regular since all its powers are also permutation matrices and hence have 0 entries.

4. We have

$$M = \begin{bmatrix} 2 & -1 \\ -17 & 9 \end{bmatrix}\begin{bmatrix} 2 & 0 \\ 0 & 3 \end{bmatrix}\begin{bmatrix} 9 & 1 \\ 17 & 2 \end{bmatrix},$$

and hence

$$M^{100} = \begin{bmatrix} 2 & -1 \\ -17 & 9 \end{bmatrix}\begin{bmatrix} 2^{100} & 0 \\ 0 & 3^{100} \end{bmatrix}\begin{bmatrix} 9 & 1 \\ 17 & 2 \end{bmatrix}.$$

5. Use a rotation matrix.

6. The matrix is
$$\begin{bmatrix} 1 & 0 \\ 0 & -1 \end{bmatrix}.$$

Eigenvectors are $[1, 0]$, with eigenvalue 1, and $[0, 1]$, with eigenvalue -1.

7. Make a change of coordinates so that the reflection is with respect to the line $y = x$.

8. We have
$$\begin{bmatrix} a & b \\ c & d \end{bmatrix} \begin{bmatrix} 1 \\ 3 \end{bmatrix} = \begin{bmatrix} 1 \\ 3 \end{bmatrix}$$

and
$$\begin{bmatrix} a & b \\ c & d \end{bmatrix} \begin{bmatrix} 3 \\ 1 \end{bmatrix} = \begin{bmatrix} 0 \\ 0 \end{bmatrix}.$$

Thus, we have a system of four equations and four unknowns. The system is easily solvable, yielding the matrix
$$\begin{bmatrix} 1/10 & 3/10 \\ -3/8 & 9/8 \end{bmatrix}.$$

The only eigenvalue is 1, and a corresponding eigenvector is $[1, 3]$.

9. The transition matrix is
$$\begin{bmatrix} 2 & -1 \\ 1 & 0 \end{bmatrix}.$$

10. $b_n = 2^n + 3^n + 5^n$, $n \geq 0$

11. Let M be the transition matrix for the Fibonacci sequence. This result is an illustration of the Cayley–Hamilton theorem, due to Arthur Cayley (1821–1895) and William Rowan Hamilton (1805–1865), which says that a square matrix M satisfies its characteristic equation.

12. (a) It is immediate that
$$p_1 = \frac{1 - p_1}{n - 1},$$

and therefore $p_1 = 1/n$.

(b) By symmetry, $p_3 = \cdots = p_n$. Let p denote the common value. Then
$$p = \frac{1}{n - 1} p_2 + \frac{1}{n - 1} (n - 3) p,$$

and hence

$$p_2 = 2p.$$

Now

$$\frac{1}{n} + 2p + (n-2)p = 1,$$

and so $p = (n-1)/n^2$. Therefore

$$[p_1, p_2, \ldots, p_n] = \left[\frac{1}{n}, \frac{2n-2}{n^2}, \frac{n-1}{n^2}, \ldots, \frac{n-1}{n^2} \right].$$

Chapter 12

◇1. Using a computer, we find that

$$\binom{n}{3}(1 - 2^{-3})^{n-3} < 1,$$

for $n = 91$. It isn't obvious how to construct a tournament with Property 3. One approach is to construct a tournament randomly on, say, 100 vertices and check to see whether the tournament has Property 3. Since

$$\binom{100}{3}(1 - 2^{-3})^{100-3} \doteq 0.38,$$

a random tournament has a good chance of having Property 3. The number of checks needed to verify that a tournament on 100 vertices has Property 3 is $\binom{100}{3} = 161700$.

◇2. Since

$$\binom{n}{12}(1 - 2^{-12})^{n-12} < 1,$$

for $n = 569459$, there exists a tournament on 569,459 vertices with Property 12.

◇3. We find that

$$\binom{n}{12}(1 - 2^{-12})^{n-12} < \frac{1}{2},$$

for $n = 572565$.

⋆4. Suppose that a tournament on seven vertices has Property 2. We claim that every vertex has outdegree three. If not, then by the pigeonhole principle some vertex x has outdegree at least four. We see that x cannot have outdegree five or six, or else a set of two vertices including x is not dominated by another vertex. Suppose that x has outdegree four, a and b are directed to x, and a is directed to b. Then the set $\{x, a\}$ is not dominated by another vertex. This contradiction shows that every vertex has outdegree three.

Next, we show that the three vertices directed to any vertex are directed in a cyclic triple. Suppose that a, b, and c are directed to x. If a, b, and c are not a cyclic triple, then they are a transitive triple, with, say, a directed to b, b to c, and a to c. But then $\{a, x\}$ is not dominated by another vertex. This establishes the claim that the three vertices directed to any vertex are directed in a cyclic triple.

Now we proceed to show that the tournament is isomorphic to the one shown in Figure 12.2. Label the three vertices to which 0 is directed 1, 2, and 4. Then the vertices 3, 5, and 6 form a cyclic triple. Choose the labeling so that 3 is directed to 5, 5 to 6, and 6 to 3. Since 3 has outdegree three, it must be directed to one of 1, 2, or 4. Choose the labeling so that 3 is directed to 4, and 1 and 2 are directed to 3. Choose the labeling of 1 and 2 so that 1 is directed to 2. We see that 5 cannot be directed to 4, because 1, 2, and 3 are not a cyclic triple. Hence 4 is directed to 5. Similarly, 1 cannot be directed to 4, because 0, 5, and 6 are not a cyclic triple. Hence 4 is directed to 1. By the same token, 4 cannot be directed to 2, because 0, 3, and 6 are not a cyclic triple. Hence, 2 is directed to 4. Continuing in this manner, we find that the remaining edges are directed just as in Figure 12.2.

†5. The result is trivially true when $n = 1$. Assume that the result holds for $n - 1$. Consider a tournament on $2^n - 1$ vertices with Property n. We will show that there exists a tournament on $2^{n-1} - 1$ vertices with Property $n-1$, a contradiction. Let v be a vertex of outdegree at least $2^{n-1} - 1$. (Such a vertex exists by the pigeonhole principle.) Consider any set of $n - 1$ vertices, not including v. Because the tournament has Property n, this set of vertices together with v is dominated by some vertex $v' \neq v$. But this means that the tournament without vertex v has Property $n - 1$.

6. Consider a directed path of maximum length. Suppose that not all the vertices are visited by the path, and derive a contradiction.

7.
$$\begin{cases} \frac{1}{24}n(n + 1)(n - 1) & \text{if } n \text{ is odd} \\ \frac{1}{24}n(n + 2)(n - 2) & \text{if } n \text{ is even} \end{cases}$$

8. Change the direction of each edge.

9. 11 of each

10. Suppose that the tournament has no Emperor and v is a King. Show that there is another King among the vertices directed to v.

†11. Try direct constructions.

12. No. Let the vertices be all integers (positive, negative, and 0), and say that a is directed to b if $a < b$.

Chapter 13

1. The number of 0s at the right of 1000! is the same as the exponent of 5 that divides 1000!, and we find that this is

$$\left\lfloor \frac{1000}{5} \right\rfloor + \left\lfloor \frac{1000}{5^2} \right\rfloor + \left\lfloor \frac{1000}{5^3} \right\rfloor + \left\lfloor \frac{1000}{5^4} \right\rfloor = 200 + 40 + 8 + 1 = 249.$$

2. 405

3. $2^{100} - 1$

4. $2^{100} - 101$

5. no

6. Since

$$\frac{(kn)!}{(n!)^k} = \binom{kn}{n, n, \ldots, n}$$

is a multinomial coefficient, it is an integer. Hence, $(kn)!$ is divisible by $(n!)^k$.

7. From the identity $\binom{n}{k} = \frac{n}{k}\binom{n-1}{k-1}$, we see that k divides $\binom{n-1}{k-1}$, since $\gcd(n, k) = 1$. Hence $\binom{n}{k}$ is a multiple of n.

8. The binary representation of $2^k - 1$ consists of all 1s; hence it dominates any number m with $0 \le m \le 2^k - 1$. It follows that 2 does not divide $\binom{2^k-1}{m}$.

9. There is a fractal-looking pattern of equilateral triangles of 0s.

10. Count the numbers whose binary expansions are dominated by the binary expansion of a given number.

11. The binary representation of $2^{10} + 2^5 + 1$ dominates the binary representation of $2^5 + 1$, so $\binom{2^{10}+2^5+1}{2^5+1}$ is not divisible by 2.

12. 5

◇13. A computer search reveals the solution $\binom{50}{3} = 140^2$.

14. Bertrand's Postulate, named after Joseph Bertrand (1822–1900), says that for any integer $n > 1$, there is a prime number between n and $2n$. It follows that $n!$ cannot be a perfect square, for there would be a prime p such that $n/2 < p < n$, and p can only divide $n!$ to the first power (since even $2p$ is greater than n).

15. We have $6! = 3!5!$, $24! = 4!23!$, and $10! = 6!7!$. The first two identities are of the form $(n!)! = n!(n! - 1)!$. It is not know whether there are solutions besides this infinite family and the third identity.

16. For $0 < i \leq j \leq n/2$, we obtain

$$\gcd\left(\binom{n}{i}, \binom{n}{j}\right) = \gcd\left(\binom{n}{i}, \binom{n}{i}\binom{n-i}{j-i}/\binom{j}{i}\right)$$

$$\geq \binom{n}{i}/\binom{j}{i}$$

$$\geq 2^i.$$

†◇17. A computer search finds the pairs $(23, 3)$ and $(90, 2)$. In fact, there is a perfect binary code corresponding to the pair $(23, 3)$ but not to the pair $(90, 2)$. See [Ple98].

Chapter 14

1. An example is $x \equiv 0 \pmod{2}$, $x \equiv 0 \pmod{3}$, and $x \equiv 1 \pmod{4}$.

◇2. Change the values of some of the k_i.

3. Infinitely many negative values of k are given in the proof of our main result. These numbers suffice in the present problem.

4. Investigate cycles of powers of 3 modulo various primes.

5. Consider the way that we defined k in the proof of our first result.

†⋆6. Use a generating function.

†7. Use Cassini's identity.

†8. Obviously $F_m \mid F_m$. Use the identity

$$F_{m+n} = F_m F_{n+1} + F_{m-1} F_n, \quad m \geq 1, n \geq 0$$

(Exercise 3 of Chapter 6).

†9. Since $\gcd(a,b) \mid a$, it follows from the previous exercise that $F_{\gcd(a,b)} \mid F_a$. Likewise, $F_{\gcd(a,b)} \mid F_b$. Hence $F_{\gcd(a,b)} \mid \gcd(F_a, F_b)$.

To prove the other direction, again use the identity

$$F_{m+n} = F_m F_{n+1} + F_{m-1} F_n, \quad m \geq 1, n \geq 0,$$

and the fact that there exist integers x and y such that $\gcd(a,b) = ax + by$.

10. The least common multiple of the m_i is 720, so we only have to check that each residue class modulo 720 is covered.

Chapter 15

1. $5, 4 + 1, 3 + 2, 3 + 1 + 1, 2 + 2 + 1, 2 + 1 + 1 + 1, 1 + 1 + 1 + 1 + 1$

   $6, 5 + 1, 4 + 2, 4 + 1 + 1, 3 + 3, 3 + 2 + 1, 3 + 1 + 1 + 1, 2 + 2 + 2,$
   $2 + 2 + 1 + 1, 2 + 1 + 1 + 1 + 1, 1 + 1 + 1 + 1 + 1 + 1$

   $7, 6 + 1, 5 + 2, 5 + 1 + 1, 4 + 3, 4 + 2 + 1, 4 + 1 + 1 + 1, 3 + 3 + 1,$
   $3 + 2 + 2, 3 + 2 + 1 + 1, 3 + 1 + 1 + 1 + 1, 2 + 2 + 2 + 1, 2 + 2 + 1 + 1 + 1,$
   $2 + 1 + 1 + 1 + 1 + 1, 1 + 1 + 1 + 1 + 1 + 1 + 1$

◇2. Do[

```
    p[i, 1] = 1; p[i] = 1;
    Do[
      If[i - j >= j,
      p[i, j] = p[i - 1, j - 1] + p[i - j, j],
      p[i, j] = p[i - 1, j - 1]];
      p[i] = p[i] + p[i, j],
      {j, 2, i - 1}
    ],
    {i, 1, 100}
    ]
```

Table[p[n, k], {n, 1, 100}, {k, 1, n}] // TableForm

Table[p[n], {n, 1, 100}] // TableForm

⋆3. The formulas are $p(n,1) = 1$, $p(n,2) = \lfloor n/2 \rfloor$, and $p(n,3) = \{n^2/12\}$. From Theorem 15.4, the generating function for the numbers $p(n,3)$ is

$$\frac{x^3}{(1-x)(1-x^2)(1-x^3)}.$$

The partial fraction decomposition of this rational function is

$$\frac{1}{6}(1-x)^{-3} - \frac{1}{4}(1-x)^{-2} - \frac{1}{72}(1-x)^{-1} - \frac{1}{8}(1+x)^{-1} + \frac{1}{9} \cdot \frac{x+2}{x^2+x+1}.$$

Expansion of the term $(x+2)/(x^2+x+1)$ reveals a repeating pattern of $-2, -1, -1, \ldots$ in the coefficients of x^n. Via the binomial series theorem and the identity $\binom{-k}{n} = (-1)^n \binom{n+k-1}{n}$, we find, by identifying the nth coefficient of the power series, that

$$p(n,3) = \frac{1}{6}\binom{n+2}{n} - \frac{1}{4}\binom{n+1}{n} - \frac{1}{72} - \frac{1}{8}(-1)^n + \frac{1}{9}C,$$

where $|C| \leq 2$. This expression simplifies to

$$p(n,3) = \frac{n^2}{12} - \frac{7}{72} - \frac{1}{8}(-1)^n + \frac{1}{9}C,$$

which is equal to $\{n^2/12\}$, since

$$\left| -\frac{7}{72} - \frac{1}{8}(-1)^n + \frac{1}{9}C \right| < \frac{1}{2}.$$

With k fixed, $p(n,k) \sim n^{k-1}/(k!(k-1)!)$. For n and k large, we can form a partition of n into k parts by writing n 1s in a row and selecting $k-1$ of them to be the rightmost terms of summation. This can be done in $\binom{n}{k-1}$ ways. Almost always, the k summands thus created will be distinct. Since the order of the parts in a partition doesn't matter, we divide by $k!$ to "unorder" the summands. Hence

$$p(n,k) \sim \frac{\binom{n}{k-1}}{k!} = \frac{n(n-1)(n-2)\ldots(n-k+2)}{k!(k-1)!} \sim \frac{n^{k-1}}{k!(k-1)!}.$$

$\diamondsuit$4. `Series[Product[(1-x^k)^-1,{k,1,20}],{x,0,20}]`

$\dagger$5. Show that the coefficients of x^n on the two sides of the equation are equal. On the right side, the product may be written as

$$\prod_{k=1}^{\infty}(1-x^k)^{-1} = \prod_{k=1}^{\infty}(1 + x^k + x^{2k} + x^{3k} + x^{4k} + \cdots).$$

6. The coefficient of x^n in the expansion of the product is the number of solutions to

$$n = m_1 + 2m_2 + 3m_3 + \cdots + nm_n,$$

where each $m_i = 0$ or 1, which is $p(n \mid \text{distinct parts})$.

7. Given a partition of n into odd parts, any two equal parts can be combined to produce a single part. Doing this as often as necessary, eventually all the parts will be distinct. On the other hand, given a partition of n with distinct parts, any even part can be split into two parts of half the length. Doing this as often as necessary, eventually all the parts will be odd.

8. By taking successive corners from a self-conjugate partition of n to build a partition of n into distinct odd numbers, we see that $\tilde{p}(n) = p(n \mid$ distinct odd parts$)$. For example, the correspondence between the self-conjugate partitions of 18 and the partitions of 18 into distinct odd parts is

$$7 + 4 + 2 + 2 + 1 + 1 + 1 \longleftrightarrow 13 + 5$$

$$5 + 4 + 4 + 4 + 1 \longleftrightarrow 9 + 5 + 3 + 1$$

$$9 + 2 + 1 + 1 + 1 + 1 + 1 + 1 + 1 \longleftrightarrow 17 + 1$$

$$8 + 3 + 2 + 1 + 1 + 1 + 1 + 1 \longleftrightarrow 15 + 3$$

$$6 + 5 + 2 + 2 + 2 + 1 \longleftrightarrow 11 + 7.$$

This correspondence yields an identity between generating functions:

$$\sum_{n=0}^{\infty} \tilde{p}(n) x^n = \prod_{n=0}^{\infty} (1 + x^{2n+1}).$$

Replacing x by $-x$, we obtain

$$\sum_{n=0}^{\infty} \tilde{p}(n)(-1)^n x^n = \prod_{n=0}^{\infty} (1 - x^{2n+1})$$

$$= \prod_{n=1}^{\infty} \frac{1}{(1 + x^n)}$$

$$= \prod_{n=1}^{\infty} (1 - x^n + x^{2n} - x^{3n} + x^{4n} - \cdots).$$

The final generating function counts

$$p(n \mid \text{even } \# \text{ of parts}) - p(n \mid \text{odd } \# \text{ of parts}).$$

9.

$$\prod_{n=1}^{\infty}\left(\frac{1}{1-x^n}-x^n\right)$$

$$=\prod_{n=1}^{\infty}\frac{1-x^n+x^{2n}}{1-x^n}$$

$$=\prod_{n=1}^{\infty}\frac{x^{3n}+1}{(1+x^n)(1-x^n)}$$

$$=\frac{(1+x^3)(1+x^6)\dots}{(1+x)(1-x)(1+x^2)(1-x^2)(1+x^3)(1-x^3)(1+x^4)(1-x^4)}\cdots$$

$$=\frac{1}{(1-x^2)(1-x^3)(1-x^4)(1-x^6)\dots}$$

10. If $m(3m-1)/2 = n(3n+1)/2$, with m and n positive integers, then $(6m-1)^2 = (6n+1)^2$, which implies that $6m-1 = 6n+1$, an impossibility.

11. We have $k(3k-1)/2 = (n(n+1)/2)/3$, upon letting $n = 3k-1$.

12.

$$9+1\longleftrightarrow 10$$
$$8+2\longleftrightarrow 7+2+1$$
$$7+3\longleftrightarrow 6+3+1$$
$$6+4\longleftrightarrow 5+4+1$$
$$4+3+2+1\longleftrightarrow 5+3+2$$

◊13.
```
p[0] = 1;
Do[
  p[n] = 0;
  k = 1;
  While[n-k(3k-1)/2 >= 0,
  p[n] = p[n] + p[n-k(3k-1)/2](-1)^(k+1); k++];
  k = 1;
  While[n-k(3k+1)/2 >= 0,
  p[n] = p[n] + p[n-k(3k+1)/2](-1)^(k+1); k++],
  {n, 1, 1000}
];
```

p[1000]

2406146786403262247369214 9727991

Chapter 16

1. 104!

◇2. The value of the feat is 26^{100}. Since this number has 142 digits and the previous number has 167 digits, the first feat has a greater value.

3. We may as well assume that the letter is given first, then the digit. The stunt is worth $\log_2(26 \cdot 10) \doteq 8.02$ bits.

4. If we learn that the event occurs, then we receive $\log_2 2^{10} = 10$ bits of information. If we learn that the event does not occur, then we receive $\log_2 1/(1 - 2^{-10}) \doteq 0.0014$ bits of information.

5. The four ways for the dice to have a sum of 9 are $(3, 6)$, $(4, 5)$, $(5, 4)$, and $(6, 3)$, and these outcomes are equally likely. We are told that both dice show an even number. Since this event occurs with probability $1/2$, the information associated with the event is $-\log_2 1/2 = 1$ bit.

6. $I(pq) = -\log_2 pq = -\log_2 p - \log_2 q = I(p) + I(q)$

7. 2 bits

8.
$$-\frac{1}{3}\log_2 \frac{1}{3} - \frac{1}{3}\log_2 \frac{1}{3} - \frac{1}{6}\log_2 \frac{1}{6} - \frac{1}{6}\log_2 \frac{1}{6} \doteq 1.9 \text{ bits}$$

9. An example is the source with states a, b, c, d, and e, with probabilities $1/2$, $1/4$, $1/8$, $1/16$, and $1/16$, respectively.

10. Since the maximum entropy of a source with n states is $\log n$, the minimum number of states necessary to have entropy 10 bits is $2^{10} = 1024$.

11. $H(T) = 1 + \frac{1}{2}H(S)$

12. Write
$$f(\alpha) = -\alpha \log \alpha - (1 - \alpha)\log(1 - \alpha),$$
where we may as well assume that the log is a natural logarithm. Then
$$f'(\alpha) = -\log \alpha + \log(1 - \alpha),$$

and the only candidate for a maximum is where $\log \alpha = \log(1 - \alpha)$, so that $\alpha = 1/2$. We can verify that this yields a maximum by checking that the second derivative of f is negative.

13. Consider Example 16.6.

14. By symmetry, $\Pr(a) = \Pr(b) = \Pr(c) = 1/3$. Hence

$$H(S) = 3(1/3)(-0.1 \log 0.1 - 0.2 \log 0.2 - 0.7 \log 0.7) \doteq 1.2.$$

15. The entropy will increase since there is no predisposition for the system to transition from a to b to c to a, as in the original Markov source.

16. Use symmetry arguments to minimize computations.

17. There is a large range of possible values for the entropy.

Chapter 17

◇1. The optimal fraction to bet is $p - q = 0.02$. The growth coefficient is

$$c = 1 + 0.51 \log 0.51 + 0.49 \log 0.49 \doteq 0.000288558.$$

At this rate, it would take $(\log_2 1000000)/c \doteq 189$ years to go from \$1 to \$1 million.

◇2. `FindRoot[0.51 Log[2, 1 + lambda]`
`+ 0.49 Log [2, 1 - lambda] == 0, {lambda, .5}]`

`{lambda -> 0.0399893}`

◇3. The optimal fraction to bet is $p - q = 0.8$. The growth coefficient is

$$c = 1 + 0.9 \log 0.9 + 0.1 \log 0.1 \doteq 0.531004.$$

At this rate, it would take $(\log_2 1000000)/c \doteq 38$ days to go from \$1 to \$1 million.

◇4. `FindRoot[0.9 Log[2, 1.0 + lambda]`
`+ 0.1 Log[2, 1.0 - lambda] == 0.0, {lambda, 0.9}]`

`{lambda -> 0.998029}`

5. Suppose that we bet λ_1, λ_2, and λ_2, respectively, on the outcomes. Of course, $p + q + r = \lambda_1 + \lambda_2 + \lambda_3 = 1$. In n trials, the amount changes on average by a multiplicative factor of

$$(\lambda_1 x)^{pn}(\lambda_2 y)^{qn}(\lambda_3 z)^{rn} = 2^{cn},$$

where

$$c = (p\log x + q\log y + r\log z) + (p\log\lambda_1 + q\log\lambda_2 + r\log\lambda_3).$$

We want to maximize the second quantity in parentheses.

By convexity (specifically, Lemma 16.4),

$$p\log\lambda_1 + q\log\lambda_2 + r\log\lambda_3 \leq -p\log p - q\log q - r\log r,$$

with equality only if $\lambda_1 = p$, $\lambda_2 = q$, and $\lambda_3 = r$.

†⋆6. Suppose that there are n outcomes, $x_1, \ldots, x_n$, with probabilities $p_1, \ldots, p_n$, respectively. We should bet p_i of the current amount on X_i, for $1 \leq i \leq n$. The proof is a straightforward generalization of the solution to the previous exercise.

7. $c(0.75) = c(0.25) = 1 + (0.75)\log_2(0.75) + (0.25)\log_2(0.25) \doteq 0.189$

8. $(0.9)^3 + 3(0.9)^2(0.1) = 0.972$

Chapter 18

1. A possible code is

$$\begin{pmatrix} a & b & c & d & e & f & g \\ p_1 & p_2 & p_3 & p_4 & p_5 & p_6 & p_7 \\ 1110 & 1111 & 100 & 110 & 101 & 00 & 01 \end{pmatrix}.$$

2. A possible code has words 0, 10, 110, 1110, ..., etc. The average length of the code is

$$\frac{1}{2}\cdot 1 + \frac{1}{4}\cdot 2 + \frac{1}{8}\cdot 3 + \cdots + \frac{1}{2^n}\cdot n.$$

It can be shown (see Chapter 3) that this sum simplifies to

$$2 - (n+2)2^{-n}.$$

3. A possible code is

$$\begin{pmatrix} a & b & c & d \\ 1/3 & 1/3 & 1/6 & 1/6 \\ 10 & 11 & 00 & 01 \end{pmatrix}.$$

The average length of the code is 2 and this is within 1 bit of the entropy, which we found to be approximately 1.9 bits.

4.

$$H(X) = -\frac{3}{4}\log\frac{3}{4} - \frac{1}{4}\log\frac{1}{4} \doteq 0.811278$$

$$H(Y) = -\frac{5}{8}\log\frac{5}{8} - \frac{3}{8}\log\frac{3}{8} \doteq 0.954434$$

$$H(X,Y) = -\frac{3}{16}\log\frac{3}{16} - \frac{9}{16}\log\frac{9}{16} - \frac{1}{32}\log\frac{1}{32} - \frac{7}{32}\log\frac{7}{32} \doteq 1.55563$$

$$H(Y|X) = -\frac{3}{16}\log\frac{1}{4} - \frac{9}{16}\log\frac{3}{4} - \frac{1}{32}\log\frac{1}{8} - \frac{7}{32}\log\frac{7}{8} \doteq 0.74435$$

$$I(X,Y) = H(Y) - H(Y|X) \doteq 0.954434 - 0.74435 = 0.210084$$

$$H(X|Y) = H(X) - I(X,Y) \doteq 0.811278 - 0.210084 = 0.601194$$

5. $(\log_2 3)/3$

6. The rate of the triplicate code is $(\log_2 2)/3 = 1/3$. In the triplicate code, an error occurs when two or three bit errors are committed (if no bit errors or only one are committed, then we can recover the intended symbol). This happens with probability

$$(0.1)^3 + 3(0.1)^2(0.9) = 0.028.$$

7. The code is a $(7, 3, 1)$ Hamming code.

†8. Center a Hamming sphere of radius $\lfloor (d-1)/2 \rfloor$ at each codeword. These spheres must be disjoint.

9. See the solution to Exercise 7.

⋆10. Use order statistics.

Chapter 19

1. 11

2. The unique self-complementary graph of order 5 is the 5-cycle C_5.

$\star$3. The "only if" assertion comes from the parity of the number of edges of the complete graph of order n. The "if" part comes from a construction. Arrange the n vertices cyclically and use for an isomorphism a cyclical shift by one vertex. Decide on the edges and non-edges accordingly.

4. By the pigeonhole principle, there exist m and n, with $m < n$, such that $13^m \equiv 13^n \pmod{10^4}$. Since $\gcd(13, 10^4) = 1$, it follows that $13^{n-m} \equiv 1 \pmod{10^4}$.

   We can find such an exponent using Euler's theorem:

   $$a^{\phi(m)} \equiv 1 \pmod{m},$$

   if $\gcd(a, m) = 1$. The furnished exponent is $\phi(10^4) = 4000$.

5. Use the same method as in the previous problem.

6. Consecutive integers are relatively prime.

7. Consider the largest power of 2 that divides each number.

8. $2n - 1$

†9. Consider the case of equality in Proof 2 of Theorem 19.2.

†$\star$10. See [Har69].

11. Color the edges of a 5-cycle C_5 green and the other edges red.

†12. Label the vertices 1 through $f(n)$ and color edge ij green if i is directed to j and red if j is directed to i.

13. Assign to each pair of vertices incident at a vertex a score of $+2$ if they are the same color and -1 if they are not the same color.

14. Let the green subgraph of K_8 be an 8-cycle with two strategically chosen diagonals.

15. Use the inequality $R(3,5) \leq R(3,4) + R(2,5) = 9 + 5 = 14$. Then find a 2-coloring of K_{13} that shows $R(3,5) > 13$.

†16. See [GRS90].

17. Give a proof similar to the one of the result in the chapter introduction.

†18. Recall that $R(a, 2) = a$ for all $a \geq 2$, and $R(a, b) \leq R(a - 1, b) + R(a, b - 1)$ for all $a, b \geq 3$.

We use induction on a and b. Note that $R(a, 2) = a = \binom{a}{a-1}$ and $R(2, b) = b = \binom{b}{b-1}$, so the inequality holds when $b = 2$ or $a = 2$. Suppose that the inequality holds for $R(a - 1, b)$ and $R(a, b - 1)$, for $a, b \geq 3$. Then

$$
\begin{aligned}
R(a, b) &\leq R(a - 1, b) + R(a, b - 1) \\
&\leq \binom{a + b - 3}{a - 2} + \binom{a + b - 3}{a - 1} \\
&= \binom{a + b - 2}{a - 1}.
\end{aligned}
$$

†19. We must show that

$$
\binom{2^{a/2}}{a} 2 < 2^{\binom{a}{2}}.
$$

Use the bound

$$
\binom{2^{a/2}}{a} = \frac{2^{a/2}(2^{a/2} - 1) \ldots (2^{a/2} - a + 1)}{a!} < \frac{(2^{a/2})^a}{2^{a/2-1}}.
$$

◇20. $100 < R(10, 10) \leq 184756$

Chapter 20

1. Since $r(C_4) = 6$, we only have to show that the second player has a drawing strategy on K_5.

2. Prove it by cases.

3. The solution does not appear to be known, so here is a chance for an independent discovery.

†4. Use mathematical induction.

⋆5. See [EH84].

6. This is another open-ended problem where you can make independent discoveries.

◇7. $n = 100$

†8. $\lfloor 2^{k/2} \rfloor$, $k \geq 3$ (see p. 161)

†9. Perhaps a computer will be of help.

†10. On a 14×15 board, there is always a winner in both the achievement and avoidance games, as follows from the Bacher–Eliahou result. The best strategy is an open question. Bacher and Eliahou constructed 14×14 binary matrices and $13 \times \infty$ binary matrices without four equal entries at the corners of a square (with horizontal and vertical sides), but the games may still be decisive on these "boards." As there are exactly 48,364 binary 14×14 matrices with an equal number of 0s and 1s and *without* the desired goal configuration, if two players play randomly on a 14×14 board, then the game ends in a draw with probability $48364/C(14^2, 14^2/2) \doteq 8.46 \times 10^{-54}$.

Chapter 21

1.
$$\begin{bmatrix} \backslash & - & | & - & / \\ | & - & - & | & | \\ - & | & \cdot & | & - \\ | & | & - & - & | \\ / & - & | & - & \backslash \end{bmatrix}$$

2. The value $k = 35$ yields a draw.

3. The animals occur in the figure as listed, reading left to right and top to bottom.

4. The first player puts an O in the center square and then reflects each of the second player's moves.

5. Elam, Domino, Tic, El, Knobby, and Elly

6. Draw a domino tiling of the plane and see which animals are blocked by it. Repeating this process until you account for all the (known) minimal non-winning animals, you will obtain diagrams similar to those following.

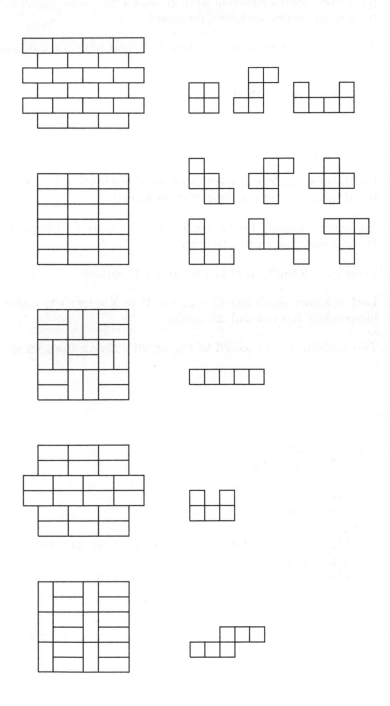

7. Play some sample games.

8. If n is even, then a reflection strategy allows the second player to win. No general results are known for n odd.

9. This problem is unsolved, so there is opportunity for independent discovery.

10. Play some sample games.

11. Play some sample games.

12. Fatty, Skinny, and Tippy

13. Use the pigeonhole principle to show that, when filled, a large enough board must contain any given Picasso animal.

14. The game is a draw. Find a domino blocking pattern. The game with "wrap-around" is a first player win.

15. Consider superanimals of non-winning 2-D animals.

16. Little is known about one-color games. Here is a chance to make some independent theories and discoveries.

17. This problem can be solved by testing all different cases up to symmetry.

Chapter 22

◇1.
```
binarystrings[n_] := (
    x = Table[0, {n}];
    Print[x];
    While[
    x != Table[1, {n}],
    place = n;
    While[x[[place]] == 1, x[[place]] = 0; place--];
    x[[place]] = 1;
    Print[x]
    ]
    )
```

2. $\{\}$, $\{1\}$, $\{1, 2\}$, $\{1, 2, 3\}$, $\{1, 2, 3, 4\}$, $\{1, 2, 4\}$, $\{1, 3\}$, $\{1, 3, 4\}$, $\{1, 4\}$, $\{2\}$, $\{2, 3\}$, $\{2, 3, 4\}$, $\{2, 4\}$, $\{3\}$, $\{3, 4\}$, $\{4\}$

◇3.
```
subsets[n_] := (
  ourlist = {{}, x = {1}};
  While[
  x != {n},
  If[x[[-1]] == n, x = Drop[x, -1]; x[[-1]]++,
    AppendTo[x, x[[-1]] + 1]];
  AppendTo[ourlist, x]
  ];
  ourlist
  )
```

4. $\{0,0,0\}, \{0,0,1\}, \{0,1,1\}, \{0,1,0\}, \{1,1,0\}, \{1,1,1\}, \{1,0,1\}, \{1,0,0\}$

◇5.
```
generalizedgraycode[n_, k_] := (
  x = Table[0, {n}];
  ourlist = {y = Table[0, {n}]};
  While[
  x != Table[k - 1, {n}],
  place = n;
  While[x[[place]] == k - 1, x[[place--]] = 0];
  x[[place]]++;
  y[[place]] = Mod[y[[place]] + 1, k];
  AppendTo[ourlist, y]
  ];
  ourlist
  )
```

◇6. Put in a check to see that the number of elements in a subset is even.

◇7.
```
graycode[1] = {{0}, {1}};
graycode[n_] :=
  graycode[n] =
    Join[Prepend[#, 0] & /@ graycode[n - 1],
      Prepend[#, 1] & /@ Reverse[graycode[n - 1]]];
```

8. A number $d_n d_{n-1} \ldots d_1 d_0$ in the ordinary listing corresponds to a number $g_n g_{n-1} \ldots g_1 g_0$ in the generalized Gray code listing if

$$g_i = d_i - d_{i+1}, \quad 0 \le i \le n,$$

and

$$d_i = g_i + \cdots + g_n, \quad 0 \le i \le n.$$

Here we have set $d_{n+1} = g_{n+1} = 0$.

To prove this, check that after a carry, only one digit advances. A carry amounts to adding a vector of the form $(0, 0, \ldots, 0, 1, 1, \ldots, 1)$, an operation that can be interpreted as finite integration. The opposite operation, finite differentiation, turns this vector into the vector $(0, 0, \ldots, 0, 1, 0, \ldots, 0,)$.

†9. A good reference for the Tower of Hanoi puzzle is the Wikipedia page `http://en.wikipedia.org/wiki/Tower_of_hanoi`.

10. Given a generalized Gray code based on n and k, the graph G has as vertices all n-tuples over $\{0, 1, \ldots, k - 1\}$, with two vertices adjacent if and only if the two n-tuples differ by 1 in exactly one coordinate. The graph G is sometimes called a *generalized hypercube*.

Chapter 23

◇1.
```
ourfactorial[n_] := n ourfactorial[n - 1];
ourfactorial[0] = 1;

ourfactorial[40]
815915283247897734345611269596115894272000000000
```

◇2.
```
f[n_, n_] := 1;
f[n_, 0] := 1;
f[n_, k_] := f[n, k] = f[n - 1, k - 1] + f[n - 1, k];

f[6,3]
20
```

3. In dictionary order, we change as little of the beginning of the word as possible. For example, ABSTRUSE appears shortly after ABSTRACT. In determining the next word, we want to leave unchanged as many letters at the left as possible. This means that we must change the rightmost letter that isn't part of a decreasing sequence of letters at the right. Since the sequence KHFCA is decreasing, and the letter to the left, E, is not part of a decreasing sequence with this sequence, the E must be changed. What letter do we change E to? Clearly, we want to pick the alphabetically least letter to the right of the E that is alphabetically greater than E. This letter is F. Hence we switch the E and F. Finally, the resulting sequence KHECA must be reversed, resulting in the desired word:

JMZORTXLBPSYWVINGDUQFACEHK.

◇4.
```
ourpermutations[n_] := (
   ourlist = {x = Range[1, n]};
```

```
While[
 x != Reverse[Range[1, n]],
 istar = n - 1;
 While[x[[istar]] > x[[istar + 1]], istar--];
 jstar = n; While[x[[jstar]] < x[[istar]], jstar--];
 x[[{istar, jstar}]] = x[[{jstar, istar}]];
 x = Join[Take[x, istar], Reverse[Drop[x, istar]]];
 AppendTo[ourlist, x]
 ];
 ourlist
 )
```

```
ourpermutations[4]
```

{{1, 2, 3, 4}, {1, 2, 4, 3}, {1, 3, 2, 4}, {1, 3, 4, 2},
{1, 4, 2, 3}, {1, 4, 3, 2}, {2, 1, 3, 4}, {2, 1, 4, 3},
{2, 3, 1, 4}, {2, 3, 4, 1}, {2, 4, 1, 3}, {2, 4, 3, 1},
{3, 1, 2, 4}, {3, 1, 4, 2}, {3, 2, 1, 4}, {3, 2, 4, 1},
{3, 4, 1, 2}, {3, 4, 2, 1}, {4, 1, 2, 3}, {4, 1, 3, 2},
{4, 2, 1, 3}, {4, 2, 3, 1}, {4, 3, 1, 2}, {4, 3, 2, 1}}

◇5.
```
ourcombinations[n_, k_] := (
    ourlist = {x = Range[1, k]};
    While[
    x != Range[n - k + 1, n],
    istar = k;
    While[x[[istar]] == n - k + istar, istar--];
    x[[istar]]++;
    Do[x[[j]] = x[[istar]] + j - istar, {j, istar + 1, k}];
    AppendTo[ourlist, x]
    ];
    ourlist
    )
```

```
ourcombinations[6, 3]
```
{{1, 2, 3}, {1, 2, 4}, {1, 2, 5}, {1, 2, 6}, {1, 3, 4},
{1, 3, 5}, {1, 3, 6}, {1, 4, 5}, {1, 4, 6}, {1, 5, 6},
{2, 3, 4}, {2, 3, 5}, {2, 3, 6}, {2, 4, 5}, {2, 4, 6},
{2, 5, 6}, {3, 4, 5}, {3, 4, 6}, {3, 5, 6}, {4, 5, 6}}

6. There are $(n!)^2$ such permutations. We can list them using two nested loops in conjunction with the Permutations Listing Algorithm.

†7. A good resource is

```
www.cut-the-knot.org/
Curriculum/Combinatorics/JohnsonTrotter.shtml+.
```

8. Try some examples, such as with $n = 3$ and $n = 4$.

Chapter 24

◇1. Set up a 729×324 binary matrix that encodes all possible ways to put a number into the Sudoku board. Take rows away depending on the givens of the puzzle. Then run the Exact Cover Algorithm.

2. 77

◇3. 576

◇4. Consider a checkerboard coloring of the board. For the computer program, include two copies of each set corresponding to placements of the animals in the box.

◇5. (a) 2

(b) 1010

(c) 2339

See [Mar91].

◇6. 8

See [Mar91].

◇7. 36, 6728

The number of tilings of a $2m \times 2n$ rectangle by dominoes is

$$4^{mn} \prod_{j=1}^{m} \prod_{k=1}^{n} \left(\cos^2 \frac{j\pi}{2m+1} + \cos^2 \frac{k\pi}{2n+1} \right).$$

◇8. This is an open problem. Notice that if the two removed squares form a domino and if this domino is in the corner, then the number of tilings is halved; if the removed domino is near the middle of the board, then the number of tilings is reduced by a much greater fraction.

◇9. 272

◇10. 1, 7, 131, 10012

◇11. 92

Appendix B

Notation

N	natural numbers, p. 3
$n!$	n factorial, p. 5
$P(n, k)$	number of permutations, p. 5
$C(n.k)$	number of combinations, p. 5
$\binom{n}{k}$	binomial coefficient, p. 6
$\binom{n}{n_1, n_2, \ldots, n_k}$	multinomial coefficient, p. 17
R	real numbers, p. 18
F_n	Fibonacci number, p. 27
$f(n) \sim g(n)$	asymptotic, p. 31
$O(g(n))$	big oh notation, p. 31
L_n	Lucas number, p. 32
$\{x\}$	nearest integer function, p. 50
$\Pr(E)$	probability of event, p. 67
$E(X)$	expected value, p. 68
$\mu(X)$	mean, p. 68
$\sigma(X)$	standard deviation, p. 69
$B(p)$	Bernoulli random variable, p. 69
$B(n, p)$	binomial random variable, p. 69

d_n	derangement number, p. 71
$d_b(n)$	sum of base-b digits, p. 97
$p(n)$	partition number, p. 109
$p(n, k)$	partition number, p. 109
$I(p)$	information function, p. 124
$H(p)$	entropy function, p. 126
$c(p)$	capacity function, p. 134
$r(C)$	rate of code, p. 148
K_n	complete graph, p. 155
$K_{m,n}$	complete bipartite graph, p. 155
C_n	cycle, p. 155
P_n	path, p. 155
$\delta(v)$	degree of vertex of graph, p. 155
G^c	graph complement, p. 156
$\alpha(G)$	independence number of graph, p. 156
$R(m, n)$	Ramsey number, p. 158
$R(a_1, \ldots, a_c)$	generalized Ramsey number, p. 159
$a(G)$	achievement number, p. 166
$\overline{a}(G)$	avoidance number, p. 166

Bibliography

[ASE92] N. Alon, J. Spencer, and P. Erdős. *The Probabilistic Method.* Wiley, New York, 1992.

[Bol79] B. Bollobás. *Graph Theory: An Introductory Course.* Cambridge University Press, New York, 1979.

[CG99] F. Chung and R. L. Graham. *Erdős on Graphs: His Legacy of Unsolved Problems.* A. K. Peters, Wellesley, second edition, 1999.

[CL96] G. Chartrand and L. Lesniak. *Graphs & Digraphs.* Chapman & Hall, New York, third edition, 1996.

[CZ05] G. Chartrand and P. Zhang. *Introduction to Graph Theory.* McGraw–Hill, New York, 2005.

[EH84] M. Erickson and F. Harary. Generalized Ramsey Theory XV: Achievement and avoidance games for bipartite graphs. In *Graph Theory Singapore 1983*, volume 1073 of *Lecture Notes in Mathematics*, pages 212–216. Springer, 1984.

[Eri96] M. J. Erickson. *Introduction to Combinatorics.* Wiley, New York, first edition, 1996.

[GRS90] R. L. Graham, B. L. Rothschild, and J. H. Spencer. *Ramsey Theory.* Wiley, New York, second edition, 1990.

[Har69] F. Harary. *Graph Theory.* Addison–Wesley, Reading, 1969.

[Hel97] L. L. Helms. *Probability Theory: With Contemporary Applications.* W. H. Freeman, New York, 1997.

[Kel56] J. L. Kelly. A new interpretation of information rate. *Bell System Technical Journal*, 35:917–926, 1956.

[Mar91] G. E. Martin. *Polyominoes: A Guide to Puzzles and Problems in Tiling.* The Mathematical Association of America, U. S. A., 1991.

[Ple98]   V. Pless. *Introduction to the Theory of Error-Correcting Codes.*
          Wiley, New York, third edition, 1998.

[Rud76]   W. Rudin. *Principles of Mathematical Analysis.* McGraw–Hill,
          Inc., New York, third edition, 1976.

[Sha48]   C. Shannon. A mathematical theory of communication. *Bell
          System Technical Journal*, 27:379–423; 623–656, 1948.

[Wes95]   D. B. West. *Introduction to Graph Theory.* Prentice Hall, Upper
          Saddle River, 1995.

Index